SCIENTIFIC DIVING TECHNIQUES

SECOND EDITION

BEST PUBLISHING COMPANY

Cover Design: Michael Gray Frick
Text Layout and Design: Michael Gray Frick

No responsibility is assumed by the Publisher or Editors for any injury and/or damage to persons or property as a matter of product liability, negligence, or otherwise, or from any use or operation of any methods, product, instructions, or ideas contained in the material herein. No suggested test or procedure should be carried out unless, in the reader's judgment, its risk is justified. Because of rapid advances in the medical sciences, we recommend that the independent verification of diagnoses and drug dosages should be made.

All rights reserved. No part of this book may be reproduced, stored in a retrieval system, or transmitted in any form or by any means, electronic, mechanical, photocopying, micro-filming, recording, or otherwise, without written permission from the publisher.

Copyright 2011 by Best Publishing Company

International Standard Book Number: 978-1-930536-68-5
Library of Congress catalog card number: 2011932110

For more information contact:
Best Publishing Company
631 US Highway 1, Suite 307
North Palm Beach, FL 33408
info@bestpub.com
Phone 561-776-6066 • 561-776-7007
Fax 561-776-4008
Website: www.bestpub.com

SCIENTIFIC DIVING TECHNIQUES

A PRACTICAL GUIDE FOR THE RESEARCH DIVER

SECOND EDITION

JOHN N. HEINE

BEST PUBLISHING COMPANY

ABOUT THE AUTHOR

JOHN HEINE IS AN EXPERIENCED SCIENTIFIC and sport diver. He has dived in many areas of the world, including the Arctic and Antarctica, temperate kelp forests, estuaries, tropical reefs, and freshwater cavern environments.

John became certified as a NAUI instructor in 1980, and as an instructor trainer and course director in 1982. He is certified as an ice diving and nitrox specialty instructor, CPR instructor, DAN O_2 instructor, and as a cavern diver. He has conducted many leadership level training programs for NAUI, and was awarded NAUI's Outstanding Service Award in 1992.

John earned a B.S. in biological sciences from the University of California at Irvine, and a master's degree in marine science from the Moss Landing Marine Laboratories at California State University. He is a past president of the American Academy of Underwater Sciences (AAUS), which awarded him the Conrad Limbaugh Memorial Award for Scientific Diving Leadership in 2007.

John taught scientific diving for 18 years at the Moss Landing Marine Laboratories, and has conducted scientific diving projects in many parts of the world.

John is also an accomplished writer and photographer. He has published scientific articles in journals such as the *Journal of Phycology*, *Journal of Experimental Marine Biology and Ecology*, *Marine Ecology Progress Series*, *Polar Biology*, and *Journal of Chemical Ecology*. His diving-related publications include books on *Cold Water Diving*, *Advanced Diving Technology and Techniques*, *Dry Suit Diving* for NAUI, *Scientific Blue Water Diving Guidelines* for the California Sea Grant Program, and articles in the *Proceedings of the American Academy of Underwater Sciences*, *NOAA Diving Manual*, *SOURCES, the Journal of Underwater Education*, and *Underwater USA*. His underwater photographs have appeared in catalogs, textbooks, magazines, and educational programs. He is a contributing editor for *SOURCES, the Journal of Underwater Education*.

John worked as the Diving Safety Officer for Moss Landing Marine Laboratories of the California State University for many years, and currently is the editor of *CalCOFI Reports*. He has also served as a member of the diving control board for the National Science Foundation, Office of Polar Programs.

ACKNOWLEDGMENTS

My introduction to scientific diving started in 1979 as a graduate student at the Moss Landing Marine Laboratories. Several courses there teach research diving methods and techniques, especially with respect to kelp forest ecology. I learned many underwater techniques through these courses. Dr. Michael Foster, emeritus professor of marine science at the Moss Landing Marine Laboratories, was an early mentor for me, as a leader in research using scientific diving.

I also learned a lot about scientific diving through participating in many research cruises and expeditions, often as a volunteer. These include Moss Landing Marine Laboratories' blue water diving cruises, the Channel Islands Research Program, a National Undersea Research Program (NURP) project at the Catalina Marine Science Center, Channel Islands National Park Service Kelp Forest Monitoring cruises, cruises to the Arctic, research trips to Antarctica, a NURP project studying deep water rockfish, a NURP project using nitrox to study deep water algae, research diving cruises in the Southern Ocean with the University of Western Australia, and the many workshops sponsored by the AAUS and the hosting university scientific diving programs.

Some of the early pioneers in scientific diving training, including Jim Stewart and Lee Somers, paved the way for scientific diving programs throughout the United States. They have always been good friends and colleagues, and have been very helpful to me throughout my career.

I would not have been able to produce this book without the generous donation of photographs that were contributed by a large number of people. These include M. Abbiati, D. Andersen, J. Auer, M. Awai, C. Bianchi, J. Bodkin, J. Bozanic, A. Buckner, Bulletin of Marine Science, E. Burge, D. Canestro, M., G. Caramanna, B. Carlson, F. Cinelli, K. Clifton, K. Coale, N. Crane, P. Cunnison, M. Dardeau, G. Davis, Q. Dokken, M. Edwards, EnviroScience, Inc., J. Estes, G. Eyal, M. Flagg, K. Flanagan, M. Foster, A. Gelber, D. Gochfeld, N. Goldberg, D. Gouge, T. Gray, S. Haddock, S. Halewood, S. Hamilton, M. Harris, E. Harvey, M. Hay, L. Hesla, W. High, D. James, Harbor Branch Oceanographic Institution, P. Iampietro, K. Joe, D. Kesling, B. Konar, J. Krug, L. Lobel, U. Kunz, M. Miller, J. Murray, C. Nelson, R. Nichols, A. Norro, W. North, Onset Corp., PBS&J, S. Painter, D. Pence, S. Pershern, G. Peterson, PISCO, M. Ponti, J. Pye, J. Reed, T. Rioux, L. Rogers-Bennett, R. Robertson, J. Rotman, K. Ruetzler, N. Schiel-Rolle, S. Sellers, B. Seymour, S. Sheldrake, H. Spalding, M. Steele, C. Storlazzi, G. Symonds, T. Thomsen, K. Trax, S. vonArbin, T. Wagner, M. Walday, N. Wu, C. Yanch, and P. Yund.

I participated as a staff member for the course/seminar for the Instructors of European Scientific Diving in Italy in 1997. There I met a number of active scientists and diving supervisors who contributed many photographs for this book. These include M. Abbiati, F. Cinelli, J. G. Harmelin, A. Norro, G. Thouzeau, and M. Walday. I thank Marco Abbiati for inviting me to participate in that course.

The American Academy of Underwater Sciences (AAUS), formed in 1977, has been instrumental in developing safe and practical scientific diving guidelines and standards. My good friend and colleague, Michael Lang, was an original organizer of that group, and has served on its board of directors for many years. He organized and produced many volumes on areas of special scientific diving significance, which have been useful in the preparation of this book. The annual Proceedings of the AAUS contain many of the original articles on scientific diving techniques that have been referenced herein.

FOREWORD

I STARTED RESEARCH DIVING IN THE LATE 1970S. At that time, the best reference books on the subject were the *NOAA Diving Manual* and Lee Somers' *Research Diver's Manual*. Much of the information in those manuals covered topics that are now found in recreational diving training and materials (i.e. physics, physiology, first aid and emergency procedures, dive planning, decompression, equipment, and the diving environment). There were no "how-to" books on the actual techniques that scientific divers use for underwater research. As in much of science, the techniques could only be found by researching the literature, or from colleagues through word of mouth.

In 1990, Jim Coyer and Jon Witman authored *The Underwater Catalog: A Guide to*

FIGURE 1:
A depiction of subtidal research in a kelp forest.
(photo courtesy of G. Davis of the National Park Service)

Methods in Underwater Research. This manual has been extremely helpful for the diving scientist looking for techniques, materials, and vendors for tools and products used in underwater research. A second edition was published by Coyer et al., in 1999.

The concept behind my book is to provide a more detailed and updated reference on scientific diving techniques. A student or researcher can read about the various techniques and methods in each chapter, and use the extensive reference sections for articles that describe techniques in greater detail. Perhaps the most valuable sections of the book, for diving safety officers who are training scientific divers, are the Training Exercises. These describe class training exercises, equipment required, and suggestions on how to conduct these dives.

The major sources for this book are scientific journal articles, the annual Proceedings of the American Academy of Underwater Sciences (AAUS), special workshop volumes, and contributions from the many colleagues who submitted photographs and descriptions of techniques that they employ in their research. There are also many references to equipment manufacturers and organizations that can be found on the internet.

The first edition of this book was published in 1999. Most of the scientific diving at that time was done on air with standard scuba. The biggest change in scientific diving since that time has been in the area of technical diving, which allows researchers access to environments and depths that were previously not attainable on air with traditional scuba equipment.

This book is dedicated to my wife, Dr. Nisse Goldberg, with whom I have spent many hundreds of hours underwater on many different science projects around the world.

TABLE OF CONTENTS

ABOUT THE AUTHOR V

ACKNOWLEDGMENTS VII

FOREWORD IX

CHAPTER 1 GENERAL SCIENTIFIC GUIDELINES 1

CHAPTER 2 AQUATIC HABITATS AND ECOSYSTEMS 19

CHAPTER 3 SPECIALIZED DIVING EQUIPMENT AND PROCEDURES 37

CHAPTER 4 LOCATING AND MARKING SITES 69

CHAPTER 5 ARCHAEOLOGY 83

CHAPTER 6 MEASURING PHYSICAL FACTORS 97

CHAPTER 7 MEASURING BIOTIC FACTORS
MATERIALS AND METHODS FOR UNDERWATER SAMPLING AND EXPERIMENTATION . . . 131

CHAPTER 8 UNDERWATER PHOTOGRAPHY/VIDEOGRAPHY FOR THE SCIENTIST . . 207

INDEX 227

TRAINING EXERCISES

1. INTRODUCTION TO SPECIALIZED EQUIPMENT AND TECHNIQUES 57

2. LOCATING AND MARKING A SITE ON SURFACE AND UNDERWATER 72

3. PRACTICE USING TOOLS UNDERWATER (MULTI-STATION EXERCISE) 73

4. GENERAL UNDERWATER MAPPING. 76

5. ARCHAEOLOGY UNDERWATER MAPPING IN POOL. 88

6. GEOLOGICAL MAPPING 101

7. MEASUREMENT OF PHYSICAL OCEANOGRAPHY PARAMETERS 115

8. SAMPLE SURVEY OF COMMON BENTHIC BIOTA: FAMILIARIZATION WITH TRANSECT TAPES, QUADRATS, AND SLATES. 144

9. FISH TRANSECTS. 149

10. COLLECTING AND TAGGING TECHNIQUES. 167

11. PHOTOGRAMETRIC TRANSECT FOR BENTHIC POPULATION ESTIMATIONS 212

12. QUANTITATIVE VIDEOGRAPHY: FISH TRANSECTS. 217

CHAPTER 1

GENERAL SCIENTIFIC DIVING GUIDELINES

Introduction

SCIENTIFIC DIVING HAS BEEN CONDUCTED in every conceivable variety of environment, including coral reefs, mangroves, kelp forests, rocky shores, soft bottom habitats, polar environments, open ocean/blue water environments, offshore platforms, estuaries, hot springs and hypersaline environments, caves, caverns, lakes, quarries, and rivers. Disciplines studied cover the range of sciences including physical, chemical, geological, biological, paleontological, and archaeological sciences.

Many innovations in diving training, equipment and design, modification, and fabrication of scientific equipment for use underwater have been developed over the past few

FIGURE 1.1:
Scientific divers conducting transect studies.
(photo courtesy of E. Burge)

decades. In fact, traditional sport scuba diving training arose from the first diver training programs developed at the Scripps Institution of Oceanography of the University of California. Scientific diving has pioneered or utilized technology such as full face masks; surface-supplied diving; tethered diving in currents, under ice, and in zero visibility conditions; the use of rebreathers, nitrox and mixed gas diving; underwater habitats; saturation diving; and the use of diver propulsion vehicles and sleds.

Scientific divers have also found many useful applications for underwater photographic and videographic cameras, research diving tools such as quadrats and transect tapes, devices for collecting, tagging, and caging organisms, data collection and recordkeeping, using epoxies and cements, drilling, lifting, and handling chemicals underwater. Many of these environments, habitats, equipment, training techniques, and underwater sampling methods will be discussed throughout this book.

As technology evolves, so also will research methods, as scientists strive to refine their techniques and efficiency underwater. However, one important thought should be remembered by all underwater researchers, as Clifton (1996) so eloquently stated: "More important than any method is the reason for its development: the need to collect data as unobtrusively and as accurately as possible."

The History of Scientific Diving

THE USE OF DIVING AS A TOOL FOR SCIENTIFIC RESEARCH first dates back to 332 B.C. when Alexander the Great was lowered into the ocean to observe marine life. Some of these early dives were probably breath-hold or bell-type diving. Typical tasks were simple collections and observations. In Sicily in 1844, Henri Milne Edwards used a commercial diving suit to make underwater investigations to a depth of 8 m (25 ft). In 1893, Dahl was probably the first person to quantitatively census a benthic community. In the early 1900s, surface-supplied ("hard hat") diving was available for a select few individuals. In 1930, scientists from the Florida Geological Survey and Smithsonian Institution used hard-hat diving equipment to collect fossil skeletons from a Florida spring (Gerrell, 1986). During the same period, Gislen pioneered underwater quantitative benthic surveys and collections. In 1934, Kitching and others made brief observations in *Laminaria* beds using a diving helmet (Kitching et al., 1934). Hans Hass used closed-circuit oxygen rebreathers for science and photography in Europe in the 1940s and 1950s.

Modern scientific diving was born in the U.S. in 1949, when Conrad Limbaugh first used a self-contained underwater breathing apparatus (SCUBA) at the Scripps Institution of Oceanography of the University of California. This allowed researchers to work directly in the underwater environment. Rather than collecting data indirectly, scientists were now able to conduct experiments underwater. In 1951, the University of California conducted the first formal scuba diving training program, and in 1953 authorized the use of scuba diving in support of science (J. Stewart, pers. comm.). One of the earliest quantitative studies using scuba was published shortly thereafter (Aleem, 1956).

After George Bond of the U.S. Navy developed saturation diving techniques allowing extended diving time underwater in the early 1960s, special underwater habitats were constructed and used for saturation diving (see Chapter 3). The first underwater habitats were inflatable bags moored to rocks on the seafloor (MacInnis, 1966; Ray, 1976). Two div-

FIGURE 1.2:
Wheeler North, an early kelp forest researcher, tying kelp transplants to a reef. (photo courtesy of P. Cunnison)

ers spent four days living in the habitat and diving daily. The SEALAB, CONSHELF, and TEKTITE projects allowed scientists the opportunity to live and work underwater for many days under saturation conditions (Somers, 1972; Collette, 1996). In 1972, the Japanese conducted an undersea saturation diving experiment in a habitat at a depth of 30 m (100 ft), then progressed to a depth of 100 m (330 ft), and planned to ultimately be able to carry out saturation dives to a depth of 300 m (990 ft) (Nakayama and Seki, 1985).

FIGURE 1.3:
Tektite II habitat from the 1960s.
(photo courtesy of W. High)

In the late 1970s, the National Oceanic and Atmospheric Administration (NOAA) established the first National Undersea Laboratory at the West Indies laboratory of Farleigh Dickinson University. This was the site of the habitat called HYDROLAB, which accomplished hundreds of scientific diving missions between 1978 and 1985 (Mastro and Dinsmore, 1987; Kesling, 1988). The name of the program was changed in the mid-1980s to the National Undersea Research Program (NURP), which currently has regional centers around the U.S., and provides support for SCUBA, nitrox, ROVs, manned submersibles, and an underwater habitat, AQUARIUS, which began operations in 1987.

FIGURE 1.4:
Aquarius underwater habitat.
(photo courtesy of M. Hay)

University Scientific Diving Programs

NEARLY EVERY COLLEGE OR UNIVERSITY THAT HAS a program in marine or aquatic sciences has a need for scientific diving capabilities. The first formal university scientific diving program in the U.S. was developed at the Scripps Institution of Oceanography at the University of California in 1954.

Large, established university marine and aquatic science programs, such as at the Scripps Institution of Oceanography, the University of California campuses, the University of Washington, and the University of Michigan, led the way in establishing training and certification protocols for scientific diving. Canadian universities formed scientific diving programs in the 1970s (Townsend, 1995).

The largest organizational member scientific diving programs in the American Academy of Underwater Sciences (AAUS) as ranked by the number of dives conducted currently are the University of California, Santa Barbara, The Florida Aquarium, Georgia Aquarium, Inc., Scripps Institution of Oceanography, the University of Hawaii, and the University of South Florida.

OSHA and the Formation of the American Academy of Underwater Sciences

IN THE LATE 1970S, A CHALLENGE TO SCIENTIFIC DIVING was made to the Occupational Safety and Health Administration (OSHA) by a labor union that contended that commercial divers could not compete for work with scientific divers due to commercial diving's extensive standards. The scientific diving community was required to prove its "case" by demonstrating an acceptable safety record achieved through good training and supervision standards. Documents submitted to OSHA included diving statistics with number of dives by depth and mode, and university and governmental scientific diving training manuals. Diving officers and scientists who used diving as a research tool testified about how scientific diving programs operate. Scientific diving was granted an exemption to the commercial diving standards in 1982, as outlined in 29 CFR Part 1910 vol. 47, no. 228, Nov. 1982, Rules and Regulations, pp. 53357-53365 (see refs.).

This battle with OSHA led to the formation of the American Academy of Underwater Sciences (AAUS; see below) in 1977, and a framework for the establishment of new scientific diving programs that would meet the criteria for exemption from the commercial diving standards was put in place. The Canadian Association for Underwater Science was founded in 1983 for many of the same reasons (Sparks, 1985; Townsend, 1995). The CAUS Standard of Practice for Scientific Diving was produced in 1985 (see http://www.caus.ca/).

Governmental and International Diving

THERE ARE A VARIETY OF SCIENTIFIC DIVING UNITS in many of the federal government agencies and departments. The National Oceanic and Atmospheric Administration (NOAA) has a large number of scientific divers, especially in the National Undersea Research Program (NURP). NOAA has over 400 divers conducting more than 15,000 dives per year. The Aquarius underwater ocean laboratory is owned by NOAA and is used by scientific divers from many academic and governmental institutions.

The National Science Foundation sponsors research around the world. The Office of Polar Programs employs a Diving Safety Officer who reviews all dive plans for scientific diving projects that occur under its auspices. This agency also has a *Guidelines for Conduct of Research Diving* manual that outlines the requirements for authorization to conduct scientific diving under the NSF/OPP program (NSF/OPP, 1993).

The Smithsonian Institution is the world's largest museum and research organization. They have ongoing research projects worldwide, where staff scientists study biodiversity, systematics, ecology, conservation, and global change processes. The Institution has research centers in Panama, Florida, Washington, D.C., Chesapeake Bay, Aldabra, and Belize.

The National Park Service has an active group of diving archaeologists, the Submerged Resources Center (formerly the Submerged Cultural Resources Unit), which uses underwater archaeology as a research and management tool (Lenihan, 1985; Bozanic, 2007; http://www.nps.gov/applications/submerged/projsub.cfm). Their mission is to enhance public appreciation and access to submerged resources. There are also numerous biologists, such as the Kelp Forest Monitoring group at the Channel Islands National Park, who perform scientific diving tasks on a regular basis. The National Biological Service of the Department of the Interior has scientific divers who work in all areas of the United States.

The University-National Oceanographic Laboratory System (UNOLS) was instituted in 1971. It consists of 62 academic member organizations that either operate or use federally funded research vessels and coordinate ship scheduling and cruises. The Research Vessel Operators' Council (now Committee; RVOC) is part of UNOLS, and is responsible for research vessel operations. Scientific diving from these vessels must operate under an AAUS-type program.

FIGURE 1.5: Shipboard diving from a large research vessel in the open ocean off Hawaii. (photo courtesy of MOBY/MLML)

Scientific diving from a large research vessel requires extensive pre-cruise planning and communications. The scientific party often comes from a number of different institutions, and a lead Diving Control Board must coordinate all dive plans and approval of personnel (Griffin and Sharkey, 1990).

The Minerals Management Service (MMS) is a federal regulatory agency within the Department of the Interior. It was formed in 1982 to improve the management of federal leasing revenues. Scientific diving activities in this group consist of gathering data for environmental impact statements, monitoring biological resources, assessing historic and prehistoric archaeological sites, and inspection for debris associated with oil and gas operations (Dauterive, 1993).

The Environmental Protection Agency (EPA) has a large scientific diving program that concentrates on environmental assessments in a variety of habitats (for example see: http://yosemite.epa.gov/r10/OEA.NSF/Investigations/Dive+BOS#eelresearch).

Figure 1.6:
EPA diver collecting sediment in a harbor.
(photo courtesy of S. Sheldrake)

The U.S. Navy has been instrumental in developing diving technology that has been incorporated into recreational and scientific diving. The U.S. Navy dive tables and the *U.S. Navy Diving Manual* are prime examples of this technology (Clarke, 2008).

Other federal agencies, such as the United States Geological Survey (USGS), the United States Forestry Service, the Federal Bureau of Investigation (FBI), and the Secret Service, have diving teams as well.

Various state and municipal agencies, such as fish and game, park services, environmental resources, water resources, and wastewater treatment facilities may also have scientific diving groups. Private agencies such as consulting firms, aquariums, and zoos may also have monitoring programs, research, or educational reasons for scientific diving.

Scientific diving in Europe is governed by groups such as the Confederation Mondiale des Activities Subaquatiques (CMAS, see http://www.cmas.org/), which was formed in 1959. In Australia, scientific diving is classified as "occupational diving," which is defined as diving performed in the course of employment, such as a part of a business, as a service, for scientific research as an educational activity, or for profit. The Australian/New Zealand Standard AS/NZ 2299.2, Occupational Diving Operations: Scientific Diving, is not a formal law, but is the standard of the community that would be used to judge scientific diving operations in these countries.

American Academy of Underwater Sciences

THE AMERICAN ACADEMY OF UNDERWATER SCIENCES (AAUS), is a professional organization of diving scientists that is dedicated to the establishment and maintenance of standards of practice for scientific diving. It is recognized as the national scientific diving entity by the Occupational Safety and Health Administration (OSHA), the National Science Foundation (NSF), the National Oceanic and Atmospheric Administration (NOAA), the University-National Oceanographic Laboratory System (UNOLS), the Underwater Hyperbaric Medical Society (UHMS), the Association of Diving Contractors (ADC), the Diving Equipment and Marketing Association (DEMA), and recreational diving agencies.

The Academy's goals are to promote the safety and welfare of its members who engage in underwater sciences. To this end, the academy provides a national forum for the exchange of information on scientific diving through annual symposia and workshops, and collects and distributes statistics on scientific diving exposure. There are various membership categories in the AAUS, including individual and organizational members. See the AAUS Web page (aaus.org) or your diving safety officer for details.

TABLE 1.1:
AAUS Organizational Members Diving Statistics

	2005	2006	2007	2008	2009
Organizational Members Reporting	92	85	94	98	98
Number of Divers	4,050	4,101	4,065	4,252	4,141
Number of Dives	126,412	126,831	124,898	122,906	113,527

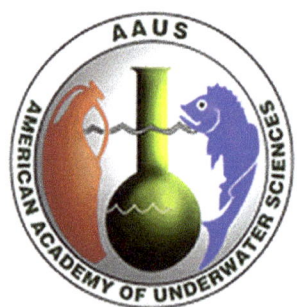

FIGURE 1.7:
The American Academy of Underwater Sciences is the recognized group for scientific diving in the US.
(logo courtesy of AAUS)

Organizational Members of the AAUS as of 2010

AECOM Design + Planning
Alaska SeaLife Center
Amegen Consulting, LLC
Aquarium of the Pacific
Arizona State University
Bermuda Institute of Ocean Science
Boston University
Broward County, FL Natural Resources Planning & Management Division
California Academy of Sciences
California Department of Fish & Game
California Science Center Foundation California State University
California State University Monterey Bay
CIEE Research Station Bonaire
Coastal & Marine Ecology Consultants, Inc.
Coastal Carolina University
Coastal Planning & Engineering, Inc.
CSA International, Inc.
Dauphin Island Sea Lab
Delta Seven Inc.
Dept. of Ecology and Evolutionary Biology, Cornell University
Duke University School of the Environment Marine Laboratory
East Carolina University
Fathom Research, LLC
Florida International University
Florida State University
Florida State University Panama City
George Mason University
Georgia Aquarium, Inc.
Glendale Community College
Harbor Branch Oceanographic Institute
Hawaii Division of Aquatic Resources
Hawaii Preparatory Academy
Humboldt State University
LUMCON
Loyola Marymount University
Magothy River Association
Marine Biological Laboratory
Marine Science Group
MBC Applied Environmental Sciences
Merkel & Associates, Inc.
Monterey Bay Aquarium
Moody Gardens
Moss Landing Marine Laboratories
Mote Marine Laboratory
National Aeronautics and Space Administration
NC Aquarium at Ft Fisher
New York Aquarium
NIWA New Zealand
Noble Odyssey Foundation
NC Aquarium at Roanoke Island
Northeastern University
Nova Southeastern Univ. Oceanogr. Center
Old Dominion University
Oregon Coast Aquarium
Oregon Health & Science University
Oregon State University
Oregon Zoo
PBS&J
Pennsylvania State University
Perry Institute for Marine Science, Caribbean Marine Research Center
Point Defiance Zoo and Aquarium
PWSSC
R. Christopher Goodwin and Associates
Richard Stockton College
Rutgers University
Saint Mary's College of California
San Diego State University
San Francisco State University/Romberg Tiburon Center
Scripps Institution of Oceanography
SCUBAnauts International, Inc.
Sea Life Park Hawaii
Seattle Aquarium
Shannon Point Marine Center
Shark Reef at Mandalay Bay
Shoals Marine Laboratory
Smithsonian Institution
South Carolina Aquarium
Southern California Ocean Restoration Divers
Stanford University
Stevens Institute of Technology
Texas A&M Galveston
Texas Parks and Wildlife Department
Texas State Aquarium
The Florida Aquarium
The Nature Conservancy Caribbean/Florida Programs
The Nature Conservancy, Hawaii Field Office
The University of Maine
The University of New Hampshire
The University of Texas at Austin
The University System of Georgia
Tierra Data Inc.
Universidad Andres Bello
Universidad de Guadalajara
University at Buffalo
University of Alaska
University of California, Los Angeles
University of California, Davis
University of California, Santa Barbara
University of California, Santa Cruz
University of Connecticut, Marine Sciences and Technology Center College of Earth, Ocean, and Environment
University of Florida
University of Guam, Marine Lab
University of Hawaii
University of Maryland College Park
University of Miami/RSMAS
University of Mississippi
University of North Carolina at Chapel Hill
University of North Carolina at Wilmington
University of Rhode Island
University of South Florida
University of Tasmania
University of the Virgin Islands
University of Washington
Virginia Institute of Marine Science
Washington State University
Woods Hole Oceanographic Institution

The AAUS *Standards for Scientific Diving Certification and Operation of Scientific Diving Programs* is a consensual diving safety manual that many scientific diving programs use as a base for their own particular diving safety manuals. Organizational members of the AAUS have their diving manuals peer reviewed by the AAUS to ensure that they meet the minimal standards.

The AAUS publishes a monthly newsletter called *The E-SLATE*, which contains information on equipment, employment, conferences, and recent publications. The organization also offers two annual scholarships, one for a master's-level student and the other for a PhD-level student. The AAUS also bestows two prestigious awards entitled the Scientific Diving Lifetime Achievement Award and the Conrad Limbaugh Memorial Award for Scientific Diving Leadership.

Diving Safety and Program Administration

THE FOREMOST CONSIDERATION FOR ANY DIVING OPERATION is safety. It is important that scientific divers not let the collection of data override the safety aspects of the dive. Proper dive planning and execution of the dive plan is critical for diving safety. Scientific divers generally have a diving supervisor assigned for each project. The diving supervisor, or lead diver should consider as a minimum the following for a scientific dive plan:

- Divers' qualifications and depth certifications. Dives should be planned around the competency of the least experienced diver.
- Emergency plan, including emergency contact information, nearest recompression chamber and medical facilities, and transportation options.
- Locations, depths, underwater times anticipated, and dive tables or computers to be used.
- Proposed work, equipment, emergency and first aid equipment, and boats to be used.
- Any hazardous conditions anticipated.

The diving safety officer (or his or her alternate) reviews and approves the dive plan, or asks for changes to be made.

It is each diver's responsibility to conduct a pre-dive safety check of all equipment to be used, as well as an evaluation of environmental conditions. A fundamental tenet of scientific diving is that each diver has the ability to refuse to dive if, in that diver's opinion, diving conditions are unfavorable or beyond the limits of their training. No scientific diver can be required to dive against their will. The ultimate responsibility for safety rests with each diver.

The Diving Control Board (DCB) consists primarily of active scientific divers in the organization, as well as administrative members. It establishes the requirements for diving certification, and acts as the official representative of the organization in matters concerning the scientific diving program. The diving safety officer (DSO) generally administers or supervises training, approval of dive plans, maintenance of diving records, equipment servicing, and ensures compliance with diving regulations.

Scientific divers generally have far more training than most recreational divers. They must be certified by a licensed physician to be medically qualified for training. After entry-level diving training, scientific divers complete additional theoretical aspects and practical training for a minimum cumulative time of 100 hours. This training includes a thorough swimming evaluation, current training in cardiopulmonary resuscitation (CPR), emergency oxygen administration, first aid for diving accidents, a written examination, and an examination of diving equipment. Theoretical aspects include such topics as collecting techniques, identification of common biota, photography, use of scientific equipment, small boat operations, and specialized diving techniques such as blue water diving, diving from large research vessels, cold water diving, and using special gas mixtures. Practical training includes a minimum of 12 open water dives in a variety of dive sites and diving conditions.

Additional examples of curriculum and practical areas of training for scientific diving include principles and activities appropriate to the intended area of scientific study, accident management, field neurological exam, dive rescue techniques, recognition of decompression illness, data gathering techniques, animal handling, animal behavior, installation of scientific apparatus, use of chemicals, site selection, site location and relocation, ecology, tagging techniques, photography, archaeology, scientific dive planning, coordination with other agencies, appropriate governmental regulations, AAUS scientific diving regulations, theoretical training in diving technology, specialized equipment to be used, diving in confined spaces, zero visibility diving, diving from research vessels, aquarium diving, polluted water diving, decompression theory and dive computers.

Other unique distinctions of scientific diving from recreational diving are the progres-

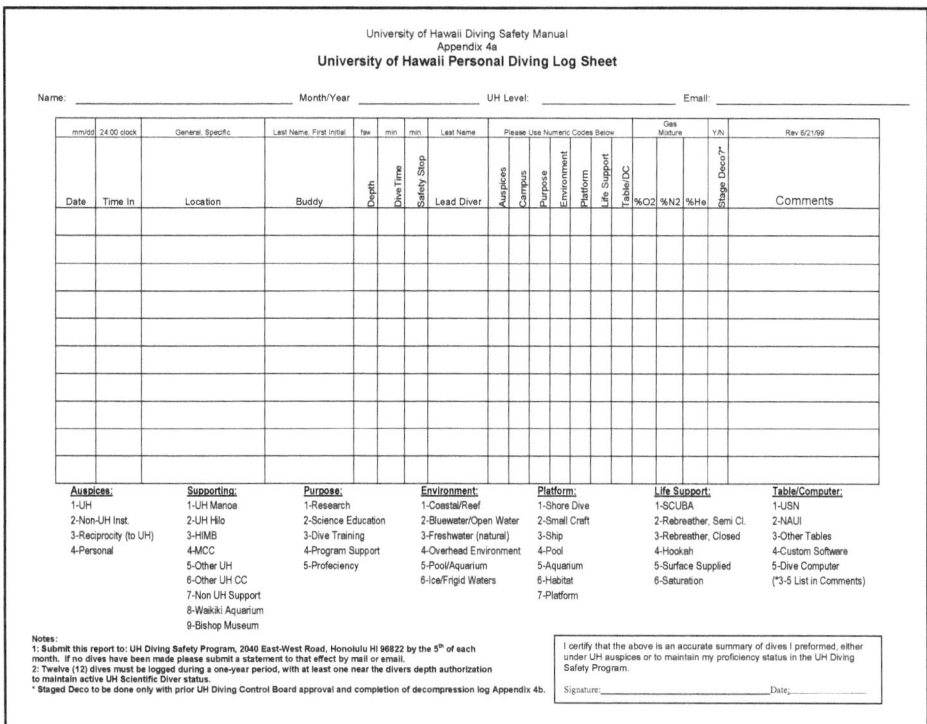

FIGURE 1.8:
Example of a scientific dive log, submitted to the DSO.

sion of depth certifications, the required submission of dive logs, continuance of scientific diving certification (which is subject to expiration), the possible revocation of diving certification, and a process for recertification. Scientific divers have a progressive step-wise depth certification procedure, which ensures adequate experience in shallow water before authorization to dive in deeper water. The general progression is from 30 feet, to 60, to 100, to 130, to 150, to a maximum depth limit of 190 feet. Divers generally are allowed to exceed their current depth certification by one step only, as long as they are accompanied by a diver certified to the next step or greater.

Scientific diving certifications are issued for a finite period of time, and are subject to expiration. For a diver to remain currently certified, a minimum of 12 dives must be logged annually, and at least one dive must be logged at or near the diver's certification depth. The AAUS has provided an electronic dive logging form that facilitates the gathering of annual scientific diving data. In addition, the medical qualification and emergency care certifications must also be current. Diving certification can be revoked by the DSO or DCB for violations of the diving safety manual, or for expiration of the items mentioned above. If a diver's certification expires, it may be renewed after complying with conditions established by the DCB or DSO.

A system of reciprocity exists between many institutions, whereby certified scientific divers can collaborate on diving projects with other researchers outside their home institution. Organizations that are members of the AAUS have agreed to adhere to the minimum diving safety standards that have been established. The AAUS has also provided a request for diving reciprocity form that verifies the diver's training and experience. The host DSO may require additional testing or demonstration of diving skills, depending upon the experience of the visiting diver, and the environment proposed to dive in.

Scientific divers must be trained to the point where the diving part of the project is almost automatic. This allows the diver to be able to concentrate on the scientific objectives without being overly concerned about things like buoyancy control, air consumption, or excessive concern for the dive buddy's well-being. Adequate thermal protection is also critical to allow the diver to make coherent decisions underwater, especially in cold water, or during long dives or periods of reduced movement or activity.

In the United States, it is important to distinguish between diving that is for science and educational purposes, and other types of diving such as commercial or military diving. This is to assure that the scientific diving being conducted meets the criteria listed by the Occupational Safety and Health Administration (OSHA) for exemption from commercial diving standards. See your diving officer if you have questions about the types of diving and techniques that you plan to use.

In general, to meet the mandate of the exemption from the commercial diving standard, a scientific diving program, project, and personnel must have these elements (Butler, 1996):

> **A Diving (Safety) Control Board consisting of a majority of active scientific divers, which has autonomous and absolute authority over the scientific diving program's operation. The DCB shall at a minimum have the authority to approve and monitor diving projects, review and revise the diving safety manual, assure compliance with**

the manual, certify the depths to which a diver has been trained, take disciplinary action for unsafe practices, and assure adherence to the buddy system (a diver is accompanied by and is in continuous contact with another diver in the water) for scuba diving.

The purpose of the project using scientific diving is the advancement of science; therefore, information and data resulting from the project are non-proprietary.

The tasks of a scientific diver are those of an observer and data gatherer. Construction and trouble-shooting tasks traditionally associated with commercial diving are not included within scientific diving.

Scientific divers, based on the nature of their activities, must use scientific expertise in studying the underwater environment and, therefore, are scientists or scientists-in-training.

A diving safety manual, which includes at a minimum the following: procedures covering all diving operations specific to the program; procedures for emergency care, recompression, and evacuation; and the criteria for diver training and certification.

For a review of the legal scope of scientific diving, with an analysis of the OSHA exemption, see Hicks (1997).

Scientific diving will always be necessary when the scientist requires a physical presence in the environment being studied. As diving technology and techniques evolve, so too will the innovations of scientific diving techniques, as researchers strive to answer questions about the aquatic environment.

Scientific Diving Contacts

American Academy of Underwater Sciences (AAUS)
Dauphin Island Sea Lab
101 Bienville Boulevard
Dauphin Island, AL 36528
Phone: (251) 591-3775
Email: aaus@disl.org
http://www.aaus.org

Australian Scientific Divers Association (ASDA)
Dr. Ed Drew
Australian Institute of Marine Sciences
PMB No. 3
Townsville, Queensland 4810
AUSTRALIA

Canadian Association for Underwater Science (CAUS)
270 Seymour River Pl.
North Vancouver, BC
Canada V7H 1W6
Phone: (604) 802-7069
Email: info@caus.ca
http://www.caus.ca/

Confederation Mondiale des Activites Subaquatiques (CMAS)
Scientific Committee
Viale Tiziano, 74
00196 Roma
Italy
Phone: 0039 06 3211 0594 / 3
Fax: 0039 06 3211 0595
Email: cmas@cmas.org
http://www.cmas.org/

The Historical Diving Society USA
2022 Cliff Dr. #119
Santa Barbara, CA 93109
Email: info@hds.org
http://www.hds.org/

Scientific Diving School
c/o Marine and Freshwater Science Group Association
Stadio Comunale Renato Dall'Ara

Implanto Carmen Longo, via Andrea Costa 174
40134 Bologna ITALY
Email: info@SDSeducational.org
http://www.sdseducational.org/

National Oceanic and Atmospheric Administration (NOAA Diving Center)
NOAA Diving Program
7600 Sand Point Way NE
Seattle, WA 98115
Phone: (206) 526-6196
Fax: (206) 526-6506
http://www.ndc.noaa.gov/

NOAA National Undersea Research Program (NURP)
1315 East-West Highway
Silver Spring, MD 20910
Phone: (301) 734-1000
Email: noaa.nurp@noaa.gov
http://www.nurp.noaa.gov/

The Online Underwater Catalog: http://tbone.biol.scarolina.edu/~dean/catalog/cover.html

University-National Oceanographic Laboratory System (UNOLS)
Graduate School of Oceanography/URI - Box 32
Ocean Science and Exploration Center
215 South Ferry Road
Narragansett, RI 02882
Phone: (401) 874-6825
http://www.unols.org/

References

Aleem, A.A. 1956. Quantitative underwater study of benthic communities inhabiting kelp beds off California. Science 123: 183.

American Academy of Underwater Sciences. 1996. Standards for Scientific Diving Certification and Operation of Scientific Diving Programs. 54 pp.

American Academy of Underwater Sciences. 2006. Standards for Scientific Diving. 80 pp.

Bozanic, J.E. 2007. An evolution of scientific mixed gas diving procedures at the National Park Services Submerged Resources Center. In: Pollock, N.W. and J.M. Godfrey (eds.), Diving for Science 2007, Proceedings of Amer. Acad. Underwater Sci. pp. 143-154.

Butler, S.S. 1996. Exclusions and exemptions from OSHA's commercial diving standards. In: Lang, M.A. and C.C. Baldwin (eds.), Methods and Techniques of Underwater Research, Proceedings of Amer. Acad. Underwater Sci. pp. 39-44.

Clarke, J.R. 2008. What can the U.S. Navy contribute to scientific diving? In: Brueggeman, P. and N.W. Pollock (eds.), Diving for Science...2008, Proceedings of Amer. Acad. Underwater Sci. pp. 37-44.

Clifton, K.E. 1996. Field methods for the behavioral study of foraging ecology and life history of herbivorous coral-reef fishes. In: Lang, M.A. and C.C. Baldwin (eds.), Methods and Techniques of Underwater Research, Proceedings of Amer. Acad. Underwater Sci. pp. 61-74.

Collette, B.B. 1996. Results of the Tektite program: Ecology of coral-reef fishes. In: Lang, M.A. and C.C. Baldwin (eds.), Methods and Techniques of Underwater Research, Proceedings of Amer. Acad. Underwater Sci. pp. 61-74.

Crayford, J.E. 1966. Underwater work. Cornell Maritime Press, Inc. Cambridge, MD. 258 pp.

Dauterive, L. 1993. Utilization of the Minerals Management Service operations and scientific diving program for operations monitoring and inspection, a case study. In: Heine, J.N. and N.L. Crane, (eds.), Diving for Science...1993, Proceedings of Amer. Acad. Underwater Sci. pp. 47-55.

Department of Labor, Occupational Safety and Health Administration (OSHA), Part 1910 of 29 CFR, Subpart T, Commercial Diving Operations. Fed. Register, July 1977, vol. 42, no. 141, pp. 37650-37676; vol. 47, no. 228, Nov. 1982, Rules and Regulations, pp. 53357-53365; and vol. 50, no. 6, Jan. 9, 1985 pp. 1046-1050.

Drew, E.A., J.N. Lythgoe, and J.D. Woods (eds.). 1976. Underwater research. Academic Press, New York. 430 pp.

Dugan, J. 1956. Man under the sea. P.F. Collier, Inc., New York.

Flemming, N.C. and M.D. Max. 1996. Scientific diving: A general code of practice. Best Publishing Co., Flagstaff, AZ. 278 pp.

Gerrell, P.R. 1986. The history and future of archaeological and paleontological work at Wakulla Springs (8WA24). In: Mitchell, C.T. (ed.), Diving for Science...86, Proceedings of Amer. Acad. Underwater Sci. pp. 59-62.

Griffin, J.J. and P. Sharkey. 1990. Workshop report: Shipboard diving safety aboard UNOLS vessels. In: W.C. Jaap, (ed.), Diving for Science...1990, Proceedings of Amer. Acad. Underwater Sci. pp. 121-137.

Haban, M.A. (ed.). 1994. Antarctic scientific diving manual. Antarctic Support Associates (ASA), Englewood, CO. 128 pp.

Haddock, S.H.D. and J.N. Heine. 2005. Scientific Blue-Water Diving Guidelines. Calif. Sea Grant Publ. No. T-057. Univ. of Calif., La Jolla, CA 92093. 49 pp.

Hamilton, R.W., D.J. Crosson, and A.W. Hulbert (eds.). 1989. Workshop on enriched air nitrox diving. National Undersea Research Program, research report No. 89-1. 153 pp.

Hanauer, E. 1996. Diving Pioneers: An oral history of diving in America. Watersport Publ., San Diego, CA.

Heine, J.N. (ed.). 1986. Blue water diving guidelines. Calif. Sea Grant Publ. no. T-CS-GCP-014. Univ. of Calif., A-032, La Jolla, CA 92093.

Heine, J.N. (ed.). 1995. Advanced diving: Technology and techniques. Nat. Assoc. Underwater Instr. (NAUI). 293 pp.

Heine, J.N. 1995. Diving dry: Skills and techniques. Nat. Assoc. Underwater Instr. (NAUI). 37 pp.

Hicks, R.E. 1997. The legal scope of "scientific diving": An analysis of the OSHA exemption. In: Maney, Jr. E.J. and C.H. Ellis, Jr. (eds.), Diving for Science...1997, Proceedings of Amer. Acad. Underwater Sci. pp. 87-100.

Kitching, J.A., T.T. Macan, and H.C. Gilson. 1934. Studies in sublittoral ecology. 1. A submarine gully in Wembury Bay, South Devon. J. Mar. Biol. Assoc. U.K. 19:677-705.

Lang, M.A. and R.W. Hamilton (eds.). 1989. Dive computer workshop. Proc. American Academy of Underwater Sciences, Univ. So. Calif. Sea Grant Publ., USCSG-TR-01-89. 231 pp.

Lang, M.A. and G.H. Egstrom (eds.). 1990. Biomechanics of safe ascents workshop. Proc. American Academy of Underwater Sciences, AAUS, AAUSDSP-BSA-01-90. 220 pp.

Lang, M.A. and R.D. Vann (eds.). 1992. Repetitive diving workshop. Proc. American Academy of Underwater Sciences, AAUS, AAUSDSP-RDW-02-92. 339 pp.

Lang, M.A. and J.R. Stewart (eds.). 1992. Polar diving workshop. Proc. American Academy of Underwater Sciences, AAUS, AAUSDSP-PDW-01-92. 100 pp.

Lang, M.A. and M.D.J. Sayer (eds.). 2007. Proc. Intl. Polar Diving Workshop. Smithsonian Institution, Wash. D.C. 213 pp.

Lang, M.A. and A.O. Brubakk (eds.). 2009. The future of diving: 100 years of Haldane and beyond. Smithsonian Institution Scholarly Press. 286 pp.

Lenihan, D.J. 1985. Underwater archeological research in the national park service. In: Mitchell, C.T. (ed.), Diving for Science...85, Proceedings of Joint International Scientific Diving Symposium, Amer. Acad. Underwater Sci. pp. 165-168.

MacInnis, J.B. 1966. Living under the sea. Scientific American, 214, no. 3: 24.

Mastro, S.J. and D.A. Dinsmore. 1987. SCUBA to submersibles, NOAA/NURP at UNC-Wilmington offers the latest technology to the marine research community. In: M.A. Lang (ed.), Coldwater Diving for Science...1987, Proceedings of the American Academy of Underwater Sciences. pp. 163-173.

Nakayama, H. and K. Seki. 1985. On the new 300m "Seatopia" project in Japan. In: Mitchell, C.T. (ed.), Diving for Science...85, Proceedings of Joint International Scientific Diving Symposium, Amer. Acad. Underwater Sci. pp. 186-189.

National Science Foundation, Office of Polar Programs. 1993. Guidelines for Conduct of Research Diving. J.R. Stewart (ed.). 18 pp.

Neushul, M. 1961. Diving in Antarctic waters. Polar Record 10(67): 353-58.

Neushul, M. 1965. Diving observations of sub-tidal Antarctic marine vegetation. Bot. Mar. 8: 234-243.

NOAA Diving Manual, diving for science and technology. 1991. U.S. Dept. of Commerce, NOAA.

NOAA Diving Manual, diving for science and technology. 2001. J. Joiner, (ed.), U.S. Dept. of Commerce, NOAA. 4th ed. Best Publ.

Peckham, V. 1964. Year-round scuba diving in the Antarctic. Polar Record 12(77): 143-146.

Ray, B. 1976. The design of a lightweight underwater habitat. In: Underwater Research, Drew, E. A., J. N. Lythgoe, and J. D. Woods (eds.), Academic Press, London, 430 pp.

Ray, C. and D. Lavallee. 1964. Self-contained diving operations in McMurdo Sound, Antarctica. Zoologica 49(8): 121-36.

Somers, L.H. 1972. Research diver's manual. The University of Michigan Sea Grant Program, tech. report no. 16, MICHU-SG-71-212.

Sparks, R.E.C. 1985. Scientific diving in Canada: CAUS activities. In: Mitchell, C.T. (ed.), Diving for Science...85, Proceedings of Joint International Scientific Diving Symposium, Amer. Acad. Underwater Sci. pp. 103-110.

Townsend, B.E. 1995. Diving regulations in Canada: Towards harmonization. In: Harper, D.E. (ed.), Diving for Science...1995, Proceedings of Amer. Acad. Underwater Sci. pp. 109-113.

Woods, J.D. and N. Lythgoe (eds.). 1971. Underwater science. Oxford University Press, London. 330 pp.

CHAPTER 2

AQUATIC HABITATS AND ECOSYSTEMS

THERE ARE A WIDE VARIETY OF AQUATIC ECOSYSTEMS on earth, most of which have been explored by scientific divers. Most aquatic environments are classified according to particular physical or biological features. Some environments are classified by the type of predominant characteristics. For example, temperate marine habitats might include an estuary, sandy beach, submarine canyon, and kelp forest all in the same bay (eg. Monterey Bay, California). This chapter will describe these environments and review scientific diving studies commonly done in them.

TROPICAL AND SUBTROPICAL SEAS

Coral reefs

CORAL REEFS OCCUR IN TROPICAL WATERS, where they can form massive geological and biological structures. They require relatively clear water with high light levels, and average water temperatures of at least 20 °C (68 °F). They can be massive geological structures made

FIGURE 2.1:
Coral reefs are found in tropical waters, and have been extensively studied by scientific divers. (photo courtesy of A. Buckner)

up of calcium carbonate that has been deposited over millions of years. Fringing reefs are the most common, and are found on rocky shores. Barrier reefs are separated from shore by a lagoon. Coral atolls are common in the Indo-Pacific, and are characterized by a coral reef ring around a central lagoon. Coral reefs are sensitive to the physical effects of sedimentation, extreme temperature, changes in salinity, and pollution.

Coral reefs can be very diverse and biologically rich, therefore much research has been conducted in these habitats (see various Proc. of Intl. Coral Reef Symposia, e.g. Gomez et al., 1982). Important studies on fish ecology (see Sale, 1991 for review), competition (Connell, 1976; Hay and Taylor, 1985); diversity (Connell, 1978; Connell, 1983), herbivory (Bakus, 1967; Hay, 1984; Hay and Fenical, 1988), climate change (Hughes et al., 2003), coral bleaching (Glynn, 1993; Brown, 1997; Hoegh-Guldberg, 1999) and fish behavior (Sale, 1972; Robertson, 1981) have been done in coral reef communities. References for research methods on coral reefs include Stoddart and Johannes (1978) and English et al. (1997).

Mangroves

MANGROVE FORESTS ARE COMMON IN TROPICAL and subtropical areas surrounding lagoons and other protected coastlines. They tend to grow in muddy waters, with long roots which branch downward underwater. These roots often harbor a diverse community of algae, invertebrates, and fishes. Mangroves are important for coastal protection (Othman, 1994) and animal habitat. There is some concern about the fate of many mangrove areas if there is a rise in sea level.

FIGURE 2.2:
Mangrove prop roots act as substratum for sponges and tunicates.
(photo courtesy of J. Reed)

FIGURE 2.3:
Rhodolith bed in Western Australia.
(photo courtesy of N. Goldberg)

Rhodolith Beds

RHODOLITH BEDS ARE COMPOSED OF UNATTACHED nongeniculate coralline algae that are found in shallow temperate and semi-tropical waters. They are widely distributed throughout the world in gently sloping soft bottoms and in channels with tidal currents, and appear to be major sources of carbonate sediments (Foster et al., 1997; Foster, 2001; Goldberg, 2006).

TEMPERATE SEAS

Soft Bottom Habitats

MUCH OF THE SUBTIDAL AREA OF THE SHALLOW marine environment has a soft sandy or muddy bottom. Although originally thought to be relatively sparsely populated, some of these environments have been shown to have dense populations of epifaunal and infaunal organisms (Dayton and Oliver, 1977). Some environments are unvegetated, while others support lush meadows of seagrasses such as eelgrass or turtle grass. These seagrass habitats are usually found in relatively protected areas and have high productivity (Hemminga and Duarte, 2000). They also harbor a diverse assemblage of associated organisms. The distribution of distinct communities is often patchy, being influenced by factors such as particle grain size and amount of organic matter present. Scientific divers have sampled these habitats using transects, quadrats, video and still photography, coring devices, air lifts, suction samplers, and dredges.

Predator-prey relationships between sea otters and bivalves have shown that otters in a California soft sediment lagoon preferred shallower-burrowing clams over larger clams, and that they had no effect on abundance and size distribution of deep-burrowing bivalves (Kvitek et al., 1988). In southeast Alaska, clams use paralytic shellfish toxins as effective deterrents to sea otter predation (Kvitek and Oliver, 1992; Kvitek, 1993; Kvitek et al., 1993). Researchers have also used side-scan sonar and diver observations of gray whale excavations of infauna on soft bottom communities in Canada (Oliver and Kvitek, 1984).

Rocky Reefs

ROCKY BOTTOM ENVIRONMENTS ARE NOT as common as soft bottom environments, but they have some of the most diverse and abundant assemblages of organisms found in the marine environment. Depending upon the location in the world, and also on the water temperature, rocky substrates may be covered with encrusting and foliose algae, as well as encrusting invertebrates, and have abundant fishes. These environments are often ideal for studying competition, predation, herbivory, succession, and recruitment events.

A considerable amount of scientific research using diving has been conducted in rocky habitats. Along the east coast of the U.S., rich algal beds of *Laminaria* sp. are common. On the west coast of north and south America, *Macrocystis* sp. forms lush kelp forests with often distinct layering and zonation depending upon depth, exposure, water clarity, light intensity, sedimentation, and storms (Foster and Schiel, 1985; Alberto et al., 2010). Dense kelp canopies have been shown to affect what species can recruit and survive under the reduced light levels (Reed and Foster, 1984). Research has also been done on sea urchin

FIGURE 2.4:
Kelp forests have been studied extensively in California since the 1950s.
(photo courtesy of J. Heine)

grazing (Harrold and Reed, 1985), sea otter/urchin/kelp interactions (Estes and Harrold, 1988; Foster and Schiel, 1988), community stability (Dayton et al., 1984), effects of El Niño (Dayton and Tegner, 1984; Bodkin et al., 1987), fish populations (Ebeling et al., 1980; Ebeling and Laur, 1988), and marine reserves (Allison et al., 1998).

Ecklonia is the dominant kelp in southern Australia and New Zealand, and considerable research has been conducted in this community as well (Choat and Ayling, 1987; Fowler-Walker and Connell, 2002).

Polar Environments (see Chapter 3)

Considerable scientific diving has been conducted in polar environments (Peckham, 1964; Neushul, 1965; Rigler, 1972; Dayton et al., 1974; Lonne, 1988; Heine, 1996; Amsler et al., 2001; McClintock and Baker, 2001). The National Science Foundation Office of Polar Programs has authorized scientific diving in polar regions since the late 1950s (Lang and Stewart, 1992; Lang and Sayer, 2007). The British Antarctic Survey (BAS) started using diving in support of its science in 1962 (White, 1995). A number of countries have research stations in Antarctica, and research vessels support diving in polar regions.

The polar environments have the most severe cold temperatures of anywhere on earth, with sub-zero air temperatures and water temperatures of -1.8 ^{0}C (29 ^{0}F) not uncommon. It is important for scientific divers to be well-trained in the use of dry suits and ice diving procedures, and to be properly outfitted for the environment (Heine, 1996).

The Arctic is characterized by ocean waters, frozen much of the year, which are surrounded by land. In contrast, the Antarctic is a frozen continent that is surrounded by

FIGURE 2.5:
Southern areas around Antarctica are characterized by rich algal beds and large sessile invertebrates.
(photo courtesy of J. Heine)

FIGURE 2.6:
In northern polar waters of the Bering Sea, sessile invertebrates are the dominant groups found on rocky substrates.
(photo courtesy of J. Heine)

ocean that is either seasonally or permanently covered with sea ice. Much of the shallow areas in polar regions are characterized by ice scour disturbance (Heine, 1989) or anchor ice disturbance (Dayton et al., 1969). Shallow rocky polar environments can support lush algal assemblages (Heine, 1989; Wynne and Heine, 1992) or extensive invertebrate communities (Dayton et al., 1974), while soft bottom areas serve as feeding grounds for migrating marine mammals.

Open Ocean, Blue Water Environment
THE PELAGIC WATERS OF THE WORLD'S OCEANS constitute the vast majority of the seas. The epipelagic zone, or upper layer, has been studied using many different means, both remote and with divers.

Scientists often need to observe or collect organisms or water samples directly, rather than remotely from a ship. For example, many gelatinous plankton communities are too fragile to collect in nets or trawls, and must be collected by divers using blue water diving techniques (Hamner, 1975; Madin and Swanberg, 1984; Biggs et al. 1986; Heine, 1986; Matsumoto, 1987; Haddock, 2003; Haddock and Heine, 2005; Gasca et al., 2007). Many studies have been conducted in this environment on feeding behavior in pteropods (Gilmer, 1972), salps (Madin, 1974) and ctenophores (Swanberg, 1974; Matsumoto, 1987); predator-prey relationships (Hamner et al., 1975; Auster et al., 1989); and on the natural history of various planktonic groups (Alldredge, 1976; Harbison et al., 1978; Purcell, 1980).

FIGURE 2.7:
Divers collecting in the blue water, open ocean environment requires special training and equipment to study.
(photo courtesy of N. Wu)

Submarine Canyons

SUBMARINE CANYONS ARE COMMON ALONG the continental shelves of many of the world's oceans, and have been explored by a few scientific divers (Shepard and Dill, 1966; Shepard, 1972; Crane and Heine, 1993; Vetter, 1998). Many of these canyons have steep walls, may have fine sediments that can obscure visibility, can have significant currents (see Inman et al., 1976), and obviously can be very deep. Much of the work done in submarine canyons is more appropriate for ROVs and manned submersibles.

Offshore Platforms

THERE ARE MANY PLATFORMS IN THE WORLD'S oceans and lakes, mainly for the purpose of drilling and extracting oil. During the 1980s, there were as many as 3,600 oil platforms in the Gulf of Mexico alone (Bull, 1989). These platforms often have extensive growth of encrusting organisms and associated fish life, creating artificial reefs in areas that would normally not support this type of growth (Wolfson et al., 1979). Addi-

FIGURE 2.8:
Diving on offshore platforms requires special planning and execution.
(photo courtesy of S. Halewood)

CHAPTER 2 AQUATIC HABITATS AND ECOSYSTEMS 25

tionally, when production ceases, the platforms must be removed, and they are often converted into artificial reefs (Bull and Kendall, 1990).

Many platforms have a maze of cables, wires, and pilings that can present problems for divers. They can also attract large fish and sharks. Diving in and around these structures requires special planning and execution.

Estuaries

ESTUARIES ARE FOUND worldwide, and are regions where freshwater from land mixes with oceanic salt water. Physical properties such as salinity, temperature, nutrient concentrations, oxygen levels, light intensity, sedimentation, and currents may all affect the types of organisms that live in estuaries. Estuaries are often rich breeding grounds for organisms that live in the ocean outside the estuary.

Diving conditions in estuaries are usually not ideal due to the sediment load and tidal action present in many of these environments. However, scientific diving has been utilized for

FIGURE 2.9:
Diver photographing rich fauna of invertebrates and fishes on offshore platform.
(photo courtesy of Q. Dokken)

physical studies of currents (Malzone, unpubl. data), collections of organisms (Hiscock, 1987), and investigations of sea otter foraging on benthic organisms (Kvitek et al., 1988). For a complete review of biological survey methods for estuaries, see Baker and Wolff (1987).

Caves and Caverns

WHILE CONSIDERABLE RECREATIONAL DIVING occurs in caves and caverns and a number of training agencies and standards exist, scientific exploration of these habitats has been minimal until relatively recently (Murphey, 1985; Murphey, 1986; Palmer, 1986; Kernagis et al., 2008). Underwater caves can be classified as freshwater, marine, or anchialine (in between the two), and many have a diverse assemblage of organisms, often morphologically, behaviorally, physiologically, or genetically different than corresponding groups in open water (Pohlman et al., 1997). Freshwater caves in Florida aquifers harbor many endemic species, some with distributions limited to a single cave system (Bauer, 1986). In the anchialine cave environment, many new species and a new order (Remipedia) of invertebrates have been discovered (Yager, 1981).

Cave and cavern environments are extremely attractive geologically and hydrologi-

FIGURE 2.10:
A freshwater spring in central Florida.
(photo courtesy of J. Heine)

cally, allowing geologists to travel far down inside the earth to collect samples. Studies have been conducted on how currents flow through aquifers, as well as how tidal action and upwelling affect mixing of fresh and salt waters (Palmer, 1986). Some interesting techniques have been used to collect data in caves, including three-dimensional mapping surveys of extensive systems, dye tracing for measuring water flows, water chemistry evaluation, and collecting of biological and geological specimens (Skiles, 1986; Schwabe, 2008).

FIGURE 2.11:
A cave diver records data on water resistant paper.
(photo courtesy of U. Kunz)

High concentrations of hydrogen sulfide (H_2S) have been reported from some cave systems, causing extreme discomfort to divers (Bozanic, 1993). Hydrogen sulfide is absorbed through the skin and burns the eyes.

FIGURE 2.12:
Freshwater environments such as the Great Lakes contain drowned forest specimens.
(photo courtesy of K. Trax)

Lakes and Rivers

A SPECIALIZED TYPE OF DIVING occurs in river environments that may have high water velocities, low visibility, cold temperatures, boat traffic, and numerous and varied types of obstructions. Areas of special concern include water withdrawls (pump intakes, siphons, or diversions), and at dams (Swan, 1988). Fishery biologists are interested in many of these withdrawl areas as a potential impact on fish populations. They may use diving to inspect and evaluate fish bypass systems for migrating fish. A diving sled has been used to survey and map salmon spawning habitats in a fast flowing and turbulent river (Swan, 1987).

At times, these environments may have zero visibility waters, which provides a challenge to the scientific diver as far as safety and scientific objectives are concerned. An ingenious zero visibility obstacle training program has been developed at East Carolina University, which prepares scientific divers to work under these conditions (Sellers and Scharf, 1990).

Stanton et al. (1994) describe diving in lakes with high concentrations of dissolved hydrogen sulfide, which is a neurotoxin. They protected divers from contact with this substance by utilizing dry suits with latex seals, hood, and dry gloves, and a full face mask with positive pressure to keep water out. They reported some minor leaking through dry suit valves, but used pieces of diaper material taped inside the suit to absorb any moisture present. Wetzel (1975) is a good reference for limnology.

FIGURE 2.13:
Some freshwater habitats are polluted and require special equipment.
(photo courtesy of EnviroScience, Inc.)

FIGURE 2.14:
Water sampling from a geothermal spring in the Mediterranean Sea.
(photo courtesy of G. Caramanna)

Hot Springs and Hypersaline Environments

SOME LAKES HAVE GEOTHERMAL FEATURES, such as hydrothermal vents, which release warm water and various chemicals and minerals. In Yellowstone Lake, scuba has been used to investigate gas fumaroles, which are surrounded by dense microbial and algal mats (Buchholz et al., 1995). Other areas of hot water may be found in areas of volcanic activity, and should be carefully evaluated by divers before approaching (Miller and Bada, 1988; Flemming and Max, 1996).

Hypersaline conditions can exist in water bodies that are subject to high evaporation, like the Dead Sea. Effects on divers include irritation of the eyes and mucous membranes. Dehydration and heat stress are likely.

Websites

http://ocean.si.edu/ocean-life-ecosystems/coral-reefs/science

http://www.mesophotic.com/

http://www.coralscience.org/main/coral-reefs

http://www.piscoweb.org/

References

Alberto, F., P.T. Raimondi, D.C. Reed, N.C. Coelho, R. Leblois, A. Whitmer, and E.A. Serrão. 2010. Habitat continuity and geographic distance predict population genetic differentiation in giant kelp. Ecology 91:49–56.

Alldredge, A.L. 1976. Field behavior and adaptive strategies of Appendicularians (Chordata: Tunicata). Mar. Biol. 38: 29-39.

Allison, G.W., J. Lubchenco and M.H. Carr. 1998. Marine reserves are necessary but not sufficient for marine conservation. Ecol. App. 8(1) Suppl.: S79-S92.

Amsler, C.D., J.B. McClintock and B.J. Baker. 2001. Secondary metabolites as mediators of trophic interactions among Antarctic marine organisms. Amer. Zool. 41(1): 17-26.

Auster, P.J., R.E. DeGoursey, and S.C. LaRosa. 1989. Observations of the interactions of gelatinous zooplankton in a nearshore environment. In: Lang, M.A. and W.C. Jaap, (eds.), Diving for Science…1989, Proceedings of Amer. Acad. Underwater Sci. pp. 1-8.

Baker, J.M. and W.J. Wolff. 1987. Biological surveys of estuaries and coasts. The Press Syndicate of the University of Cambridge. 453 pp.

Bakus, G.J. 1967. The feeding habits of fishes and primary production at Eniwetok, Marshall Islands. Micronesia 3: 135-149.

Bauer, J. 1986. Cave diving as a technique for inventorying Florida cave faunas. In: Mitchell, C.T. (ed.), Diving for Science…86, Proceedings of Amer. Acad. Underwater Sci. pp. 91-95.

Biggs, D.C., P. Laval, J-C. Braconnot, C. Carre, J. Goy, M. Masson, and P. Morand. 1986. In: Mitchell, C.T. (ed.), Diving for Science…86, Proceedings of Amer. Acad. Underwater Sci. pp. 153-161.

Bodkin, J.L., G.R. VanBlaricom, and R.J. Jameson. 1987. Mortalities of kelp forest fishes associated with large oceanic waves off central California, 1982-83. Environ. Biol. Fishes 18: 73-76.

Brown, B.E. 1997. Coral bleaching: causes and consequences. Coral Reefs 16, Suppl. 1: S129-S138.

Buchholz, L.A., P.D. Anderson, R.L. Cuhel, J.V. Klump, J.E. Kostka, R.W. Paddock, and C.C. Remsen. 1995. Employment of ROV techniques and scuba in Yellowstone lake. In: Harper, D.E. (ed.), Diving for Science…1995, Proceedings of Amer. Acad. Underwater Sci. pp. 1-7.

Bull, A.S. and J.J. Kendall. 1990. Mechanisms of outer continental shelf (OCS) oil and gas platforms as artificial reefs in the Gulf of Mexico. In: W.C. Jaap, (ed.), Diving for Science…1990, Proceedings of Amer. Acad. Underwater Sci. pp. 21-37.

Bull, D. 1989. Offshore oil platforms in the Gulf of Mexico, conversion to artificial reefs: an opportunity for long term biological studies. In: Lang, M.A. and W.C. Jaap, (eds.), Diving for Science…1989, Proceedings of Amer. Acad. Underwater Sci. pp. 25-28.

Choat, J.H. and A.M. Ayling. 1987. The relationship between habitat structure and fish faunas on New Zealand reefs. J. Exp. Mar. Biol. Ecol. 110: 257-284.

Connell, J.H. 1976. Competitive interactions and the species diversity of corals. In: Mackie, G.O. (ed.), Third Intl. Symp. on Coelenterate Biology, Univ. Victoria, Canada.

Connell, J.H. 1978. Diversity in tropical rain forests and coral reefs. Science 199: 1302-1310.

Connell, J.H. 1983. On the prevalence and relative importance of interspecific competition: Evidence from field experiments. Am. Nat. 122: 661-696.

Crane, N.L. and J.N. Heine. 1993. Observations of the prickly shark (*Echinorhinus cookei*) in Monterey Bay, California. Calif. Fish and Game 78(4): 166-168.

Dayton, P.K., G.A. Robilliard, and A.L. DeVries. 1969. Anchor ice formation in McMurdo Sound, Antarctica, and its biological effects. Science 163: 273-274.

Dayton, P.K. and J.S. Oliver. 1977. Antarctic soft-bottom benthos in oligotrophic and eutrophic environments. Science 197: 55-58.

Dayton, P.K., G.A. Robilliard, R.T. Paine, and L.B. Dayton. 1974. Biological accommodation in the benthic community at McMurdo Sound, Antarctica. Ecol. Monogr. 44: 105-128.

Dayton, P.K., V. Currie, T. Gerrodette, B.D. Keller, R. Rosenthal, and D. venTresca. 1984. Patch dynamics and stability of some California kelp communities. Ecol. Monogr. 54: 253-289.

Dayton, P.K. and M.J. Tegner. 1984. Catastrophic storms, El Nino, and patch stability in a southern California kelp community. Science 224: 283-285.

Ebeling, A.W., R. J. Larson, W.S. Alevizon, and R.N. Bray. 1980. Annual variability of reef-fish assemblages in kelp forests off Santa Barbara, California. Fish. Bull. 78: 361-377.

Ebeling, A.W. and D.R. Laur. 1988. Fish populations in kelp forests without sea otters: Effects of severe storm damage and destructive sea urchin grazing. Ecol. Stud. 65: 169-191.

English, S., C. Wilkinson, and V. Baker. 1997. Survey manual for tropical marine resources. Australian Institute of Marine Science, Townsville, Australia. 390 pp.

Estes, J.A. and C. Harrold. 1988. Sea otters, sea urchins, and kelp beds: Some questions of scale. Ecol. Stud. 65: 116-150.

Foster, M.S. and D.R. Schiel. 1985. The ecology of giant kelp forests in California: A community profile. U.S. Fish. Wildl. Serv. Biol. Rep. 85 (7.2). 152 pp.

Foster, M.S. and D.R. Schiel. 1988. Kelp communities and sea otters: Keystone species or just another brick in the wall? Ecol. Stud. 65: 92-115.

Foster, M.S., R. Riosmena-Rodriguez, D.L. Steller, and W.J. Woelkerling. 1997. Living rhodolith beds in the Gulf of California and their implications for paleoenvironmental interpretation. Geol. Soc. Amer. 318: 127-139.

Foster, M.S. 2001. Rhodoliths: between rocks and soft places. Journal of Phycology 37: 659-667.

Fowler-Walker, M.J. and S.D. Connell. 2002. Opposing states of subtidal habitat across temperate Australia: consistency and predictability in kelp canopy-benthic associations. Mar. Ecol. Prog. Ser. 240: 49-56.

Gasca, R., E. suarez-Morales and S.H.D. Haddock. 2007. Symbiotic associations between crustaceans and gelatinous zooplankton in deep and surface waters off California. Mar. Biol. 151 (1): 233-242.

Gilmer, R.W. 1972. Free-floating mucus webs: A novel feeding adaptation for the open ocean. Science 177: 1239-1240.

Glynn, P.W. 1993. Coral reef bleaching: ecological perspectives. Coral Reefs 12: 1-17.

Goldberg, N. 2006. Age estimates and description of rhodoliths from Esperance Bay, Western Australia. J. Mar. Biol. Ass. U.K. 86: 1291-1296.

Gomez, E.D., C.E. Birkeland, R.W. Buddemeier, R.E. Johannes, J.A. Marsh, Jr., and R.T. Tsuda (eds.). 1982. Proceedings of the Fourth International Coral Reef Symposium, Mar. Sci. Center, Univ. of the Philippines. 725 pp.

Haddock, S.H.D. 2003. A golden age of gelata: past and future research on planktonic ctenophores and cnidarians. Hydrobiologia 530-531 (1-3): 549-556.

Haddock, S.H.D. and J.N. Heine. 2005. Scientific Blue-Water Diving Guidelines. Calif. Sea Grant Publ. No. T-057. Univ. of Calif., La Jolla, CA 92093. 49 pp.

Hamner, W.M. 1975. Underwater observations of blue-water plankton: Logistics, techniques, and safety procedures for divers at sea. Limnol. Oceanogr. 20: 1045- 1051.

Hamner, W.M., L.P. Madin, A.L. Alldredge, R.W. Gilmer, and P.P. Hamner. 1975. Underwater observations of gelatinous zooplankton: Sampling problems, feeding biology, and behavior. Limnol. Oceanogr. 20: 907-917.

Harbison, G.R., L.P. Madin, and N.R. Swanberg. 1978. On the natural history and distribution of oceanic ctenophores. Deep-Sea Research 25: 233-256.

Harrold, C. and D.C. Reed. 1985. Food availability, sea urchin grazing, and kelp forest community structure. Ecology 66: 1160-1169.

Hay, M.E. 1984. Patterns of fish and urchin grazing on Caribbean coral reefs: Are previous results typical? Ecology 65: 446-454.

Hay, M.E. and W. Fenical. 1988. Marine plant-herbivore interactions: The ecology of chemical defense. Annu. Rev. Ecol. Syst. 19: 111-145.

Hay, M.E. and P.R. Taylor. 1985. Competition between herbivorous fishes and urchins on Caribbean reefs. Oecologia 65: 591-598.

Heine, J. N., (ed.) 1986. Blue Water Diving Guidelines. Calif. Sea Grant Program, No. T-CSGCP-014, 46 pp.

Heine, J.N. 1989. Effects of ice scour on the structure of sublittoral marine algal assemblages of St. Lawrence and St. Matthews Islands, Alaska. Mar. Ecol. Progr. Ser. 52: 253-260.

Heine, J.N. 1996. Cold Water Diving: A Guide to Ice Diving. Best Publishing. 127 pp.

Hemminga, M. and C. Duarte. 2000. Seagrass ecology. Cambridge University Press. 303 pp.

Hiscock, K. 1987. Subtidal rock and shallow sediments using diving. Biological Surveys of Estuaries and Coasts. Cambridge University Press New York. pp. 198-237.

Hoegh-Guldberg, O. 1999. Climate change, coral bleaching and the future of the world's coral reefs. Mar. and Freshw. Res. 50(8): 839-866.

Hughes, T.P., A.H. Baird, D.R. Bellwood, M. Card, S.R. Connolly, C. Folke, R. Grosberg, O. Hoegh-Guldberg, J.B.C. Jackson, J. Kleypas, J.M. Lough, P. Marshall, M. Nystrom, S. R. Plumbi, J.M. Pandolfi, B. Rosen, and J. Roughgarden. 2003. Climate change, human impacts, and the resilience of coral reefs. Science 15 (301) no. 5635: 929-933.

Inman, D.L., C.E. Nordstrom and R.E. Flick. Currents in submarine canyons: an air-sea-land interaction. Annu. Rev. Fluid Mech. 8: 275-310.

Kernagis, D.N., C. McKinlay and T.R. Kincaid. 2008. Dive logistics of the Turner to Wakulla Cave traverse. In: Brueggeman, P. and N.W. Pollock (eds.), Diving for Science 2008, Proceedings of the American Academy of Underwater Sciences. pp. 91-102.

Klos, E., J.H. Costello, S.P. Colin, and W.M. Graham. 2005. In: Godfrey, J.M. and S.E. Shumway (eds.), Diving for Science 2005, Proceedings of the American Academy of Underwater Sciences. pp. 211-216.

Kvitek, R.G. 1993. Paralytic shellfish toxins as a chemical defense in the butter clam (*Saxidomus giganteus*). Dev. Mar. Biol. 3: 407-412.

Kvitek, R.G., A.K. Fukuyama, B.S. Anderson, and B.K. Grimm. 1988. Sea otter foraging on deep-burrowing bivalves in a California coastal lagoon. Mar. Biol. 98(2): 157-167.

Kvitek, R.G. and J.S. Oliver. 1992. Influence of sea otters on soft-bottom prey communities in southeast Alaska. Mar. Ecol. Prog. Ser. 82: 103-113.

Kvitek, R.G., C.E. Bowlby, and M. Staedler. 1993. Diet and foraging behavior of sea otters in southeast Alaska. Mar. Mamm. Sci. 9(2): 168-181.

Lang, M.A. and J.R. Stewart (eds.). 1992. Polar Diving Workshop. Proceedings of the American Academy of Underwater Sciences. 100 pp.

Lang, M.A. and M.D.J. Sayer (eds.). 2007. Proc. Intl. Polar Diving Workshop. Smithsonian Institution, Wash. D.C. 213 pp.

Lonne, O.J. 1988. A diver-operated electric suction sampler for sympagic (= under-ice) invertebrates. Polar Research vol. 6 (1): 135-136.

Madin, L.P. 1974. Field observations on the feeding behavior of salps (Tunicata: Thaliacea). Mar. Biol. 24: 143-147.

Matsumoto, G.I. 1987. Manipulation of water flow by ctenophores (Phylum Ctenophora). In: M.A. Lang (ed.), Coldwater Diving for Science...1987, Proceedings of the American Academy of Underwater Sciences. pp. 151-161.

McClintock, J.B. and B.J. Baker (eds.). 2001. Marine Chemical Ecology. CRC Press, 593 pp.

Miller, S.L. and J.L. Bada. 1988. Submarine Hot Springs and the origin of life. Nature 334: 609-611.

Morris, R., R. Berthold and N. Cabrol. 2007. Diving at extreme altitude: dive planning and execution during the 2006 High Lakes Science Expedition. In: Pollock, N.W. and J.M. Godfrey (eds.), Diving for Science 2007, Proceedings of the American Academy of Underwater Sciences. pp. 155-168.

Murphey, M. 1985. The overhead diving environment: Standards. In: Mitchell, C.T. (ed.), Diving for Science...85, Proceedings of Joint International Scientific Diving Symposium, Amer. Acad. Underwater Sci. pp. 295-311.

Murphey, M. 1986. Proposed AAUS overhead environment diving standards. In: Mitchell, C.T. (ed.), Diving for Science...86, Proceedings of Amer. Acad. Underwater Sci. pp. 47-54.

Neushul, M. 1965. Diving observations of sub-tidal Antarctic marine vegetation. Bot. Mar. 8: 234-243.

Oliver, J.S. and R.G. Kvitek. 1984. Side scan sonar records and diver observations of the gray whale (*Eschrichtius robustus*) feeding grounds. Biol. Bull. Mar. Biol. Lab. Woods Hole. 167(1): 264-269.

Othman, M.A. 1994. Value of mangroves in coastal protection. Hydrobiologia 285: 277-282.

Palmer, R. J. 1986. Expedition cave diving: a new scientific tool. In: Mitchell, C.T. (ed.), Diving for Science...86, Proceedings of Amer. Acad. Underwater Sci. pp. 55-58.

Peckham, V. 1964. Year-round scuba diving in the Antarctic. Polar Record 12(77): 143-146.

Pohlman, J.W., T.M. Iliffe and L.A. Cifuentes. 1997. A stable isotope study of organic cycling and the ecology of an anchialine cave ecosystem. Mar. Ecol. Prog. Ser. 155: 17-27.

Purcell, J.E. 1980. Influence of siphonophore behavior upon their natural diets: evidence for aggressive mimicry. Science 209: 1045-1047.

Reed, D.C. and M.S. Foster. 1984. The effects of canopy shading on algal recruitment and growth in a giant kelp forest. Ecology 65: 937-948.

Rigler, F.H. 1972. Director's review. In: Char Lake Project Annual Report 1971-72. Canadian Committee International Biological Program.

Robertson, D.R. 1981. The social and mating systems of two labrid fishes, *Halichoeres maculipinna* and *H. garnoti*, off the Caribbean and Panama. Mar. Biol. 64: 327- 340.

Sale, P.F. 1972. Effect of cover on agonistic behavior of a reef fish: A possible spacing mechanism. Ecology 53: 753-758.

Sale, P.F. 1991. The ecology of fishes on coral reefs. Academic Press, Calif. 754 pp.

Schwabe, S.J. 2008. The difficulties of sampling in underwater caves in the Bahamas: an exercise in ingenuity and survival. In: Brueggeman, P. and N.W. Pollock (eds.), Diving for Science 2008, Proceedings of the American Academy of Underwater Sciences. pp. 147-158.

Sellers, S. and R. Scharf. 1990. Training scientific divers for zero visibility diving. In: W.C. Jaap, (ed.), Diving for Science...1990, Proceedings of Amer. Acad. Underwater Sci. pp. 313-322.

Shepard, F.P. and R.F. Dill. 1966. Submarine canyons and other sea valleys. Rand McNally Press, Chicago. 381 pp.

Shepard, F.P. 1972. Submarine canyons. Earth-Science Reviews vol. 8 (1): 1-12.

Skiles, W. C. 1986. The scientific future of cave diving. In: Mitchell, C.T. (ed.), Diving for Science...86, Proceedings of Amer. Acad. Underwater Sci. pp. 37-46.

Stanton, G., W. Burnett, and B. Gilam. 1994. Polluted water technology used in a miramictic lake in Belau, Micronesia. In: Petrecca, R. (ed.), Diving for Science, 1994, Proceedings of Amer. Acad. Underwater Sci. pp. 81-91.

Stoddart, D.R. and R.E. Johannes. 1978. Coral Reefs: research methods. UNESCO, Paris, 581 pp.

Swan, G.A. 1987. Use of a diver's sled and laser locating system for salmon spawning surveys. In: M.A. Lang (ed.), Coldwater Diving for Science...1987, Proceedings of the American Academy of Underwater Sciences. pp. 265-277.

Swan, G.A. 1988. Safety considerations for diving at water withdrawls and dams. In: M.A. Lang (ed.), Advances in Underwater Science...1988, Proceedings of Amer. Acad. Underwater Sci. pp. 185-205.

Swanberg, N. 1974. The feeding behavior of *Beroe ovata*. Mar. Biol. 24: 69-76.

Vetter, E.W. 1998. Population dynamics of a dense assemblage of marine detritivores. J. Exp. Mar. Biol. Ecol. 226: 131-161.

Wetzel, R.G. 1975. Limnology. Saunders College Publ., Philadelphia, PA. 743 pp.

White, M.G. 1995. Scientific diving by British Antarctic Survey: 1962-1995. In: Harper, D.E. (ed.), Diving for Science...1995, Proceedings of Amer. Acad. Underwater Sci. pp. 137-143.

Wolfson, A., G. VanBlaricom, N. Davis and G.S. Lewbel. 1979. The marine life of an offshore oil platform. Mar. Ecol. Prog. Ser. 1: 81-89.

Wynne, M.J. and J.N. Heine. 1992. Collections of marine red algae from St. Matthew and St. Lawrence Islands, the Bering Sea. Nova Hedwigia 55: 55-97.

Yager, J. 1981. Remipedia, a new class of crustacea from a marine cave in the Bahamas. J. Crust. Biol. 1(3): 328-333.

CHAPTER 3

SPECIALIZED DIVING EQUIPMENT AND PROCEDURES

OPEN CIRCUIT SCUBA DIVING IS BY FAR the most prevalent mode of scientific diving, due primarily to its wide availability, relatively light weight and portability, relative ease in training, and cost. This mode of scientific diving is limited to a maximum depth of 190 fsw (60 m) in the United States. Many other modes of diving may be desirable or required in certain environmental situations and will be described below. This is not meant to be an exhaustive list of diving modes, but merely an introduction to other modes of diving used by scientific divers. As with all types of diving, additional training is required for modes other than scuba. See your diving safety officer for details (AAUS Standards, 2006).

Scientific divers have made some adjustments and additions to standard scuba equipment over the years. Some recommendations that apply to all scuba divers include streamlining equipment to reduce dangling hoses, gauges, and straps, having adequate D-rings, clips, and snaps to hold collection bags and tools, and thorough knowledge of all equip-

FIGURE 3.1:
Scientific divers use a variety of specialized equipment such as full face masks, communications, and HD cameras.
(photo courtesy of B. Seymour)

ment, so that diving becomes "second nature" (Paulet and McLean, 1989). Some scientific divers prefer to be weighted to be slightly negatively buoyant to allow the diver to stay on the bottom in surgy conditions. For this reason, scientific divers often wear heavy-duty kneepads and strong gloves.

The importance of staying warm cannot be stressed enough. Studies have shown that divers will often not perceive that they are cold, which can lead to a degradation of mental performance (Bachrach, 1985; Ronnestad et al., 1987; Lang and Stewart, 1992; Lang and Sayer, 2007; Stinton, 2007). This can have obvious effects on diver safety and in the collection of valid data underwater. Adequate thermal protection with a wet suit, dry suit, or hot water suit is essential for the working scientific diver (Ronnestad et al., 1987).

Regulators

WHILE VIRTUALLY ANY SINGLE-HOSE SCUBA REGULATOR can be used in most scientific diving locations and applications, there are certain environments that require a careful evaluation and choice of regulators. One such environment is in cold or near-freezing water, where the first stage of many single-hose regulators will freeze up, free flow, and generally not perform adequately. A thorough review of regulator performance in Antarctic diving shows that the Sherwood Maximus and Poseidon Odin regulators performed the best of 16 regulator models (Bozanic and Mastro, 1992; Mastro and Pollock, 1995). In a more recent study, the Poseidon Xstreme performed very well in demanding freezing conditions (Lang and Sayer, 2007). The U.S. Navy has done extensive testing of regulators in cold water (Clarke, 2007). If the diver is working hard, a high-performance regulator should be chosen. In other situations, such as high sediment in the water column, a diaphragm first stage regulator might also be more appropriate.

Full Face Masks

FULL FACE MASKS OR DIVING helmets are useful when diving in polluted water, or when certain types of communications are required. A number of manufacturers produce full face masks, and popular models include

FIGURE 3.2:
The Divator/AGA full face mask is relatively economical and easy to use for communication or protection from polluted water.
(photo courtesy of B. Seymour)

Kirby Morgan Dive Systems EXO-BR and M-48 SuperMask, and the Divator AGA MK-II distributed by Interspiro. Many of these masks can be used with standard scuba or with surface supplied diving. Some useful features include positive pressure in the mask, which helps to keep any water from entering the mask, automatic defogging of the face port, and the lack of a regulator mouthpiece to bite on, which allows for easy communications and reduced jaw fatigue. Some disadvantages include the potential of using more air than with traditional scuba, and that in an out of air or mechanical failure situation, one cannot share air or easily use an octopus without removing the full face mask (Barsky, 1990; Sheldrake et al., 2009; Barsky, 2010).

Surface-Supplied Diving

SURFACE-SUPPLIED DIVING IS A MODE WHERE the diver's breathing gas and hard wire communications are supplied from the surface. The advantages of surface supplied diving include potentially unlimited breathing gas supply from a surface compressor, and/or banks of cylinders, and surface personnel controlling the diver's depth, decompression, and breathing gas supply, freeing the scientist to concentrate on the scientific tasks. Data can be spoken to the surface where it can be recorded by surface personnel or on tape. Diver to diver communications can be performed as well. Surface-supplied systems, however, are bulkier, take up more space, have a higher cost, require extra personnel on the surface, have higher maintenance costs, and the diver can't move around as much underwater due to the umbilical hose leading to the surface. Also, diving helmets can be very heavy, especially out of the water.

The basic elements of a surface-supplied diving system include a full face or band mask, or lightweight helmet, an umbilical hose which carries the breathing gas, a pneumo hose (for depth determination), communication wires from the surface console to the diver, and a low pressure compressor or high pressure gas source. The minimum team required is one diver, a surface tender, and a stand-by diver (Somers, 1987).

For diving in contaminated water, a dry suit would also be used with surface supplied diving. Diving in water that is contaminated by chemicals, biological agents, or even radioactive elements or compounds is extremely hazardous, and should only be attempted by those trained in the proper safety procedures (Phoel, 1981; Colwell, 1982; Barsky, 1986; Barsky, 1990). Stanton et al., (1994) describe diving technology used to dive in lakes that have high levels of hydrogen sulfide.

FIGURE 3.3: Kirby Morgan Dive Systems band mask with surface-supplied air and communications.
(photo courtesy of J. Heine)

FIGURE 3.4:
Diving in contaminated water requires special equipment and decontamination procedures. (photo courtesy of S. Sheldrake)

Hookah

SOME RESEARCHERS HAVE USED HOOKAH, or a modification of a shallow-water surface-supplied diving system that has the intermediate or high pressure air source on the surface. Small, portable gasoline-powered low-pressure compressors can be placed on a small boat, or even a floating inner-tube (Brownie's Third Lung), and supply one or two divers with air via low pressure hoses to a second stage regulator. One of the main advantages of hookah is that the diver does not have to wear a large scuba cylinder, increasing the time spent underwater based on air supply. Depth and mobility are limited by the length of the hose on the unit. Another advantage to this system is its relatively lightweight and long-lasting air supply, which is a consideration in remote areas where high pressure air compressors may not be available or practical (see Stanton et al., 1994).

A modification of traditional hookah has been described by Marelli and Jaap (1990), where a high pressure air cylinder with a traditional scuba first stage regulator is on the surface, connected to a long (30 m; 100 ft), lightweight intermediate hose with a second stage regulator on the diver's end. Another type of hookah is described by Walker and Gurney (1985) for recovering diamonds from the surf zone in South Africa.

Communications

IT IS OFTEN NECESSARY TO COMMUNICATE WITH a dive buddy or surface personnel by means other than hand signals, written communications, or tugs on a line. Wireless communication systems are available in a variety of different configurations. One simple and inexpensive system is called the Buddy Phone. The system includes a half-mask which connects to a standard second stage regulator, or a full face mask and a transmitter/receiver which fits on the mask strap. Various configurations will allow for diver to diver, diver to surface, and surface to diver communications. The range is from 50 to 500 meters (165 to 1650 ft), and it has a maximum operating depth of 40 m (130 ft). Half-mask configurations are somewhat bulky, and some practice is required to achieve clear speech.

Communications with full face masks can utilize single sideband technology, such as that manufactured by Ocean Technology Systems (OTS) or Scubaphone. A microphone fits in the full face mask, an earphone hangs on the mask strap, and a transceiver attaches to the diver's buoyancy compensator or on top of the head (Murdoch, 1987). This system features push to talk or voice activated communications between divers and the surface, with some models boasting a communication range of 1350 m (4,500 ft). Digital signal processing units assure clean speech and long range (up to 3,000 meters; 9,900 ft), with a variety of channels to choose from. Hard wire systems are also available, which provide a physical link to the surface (Barsky, 1990). Tether lines with communications wire inside are useful in tethered-diver operations.

FIGURE 3.5:
Surface check of communications with a diver with a full face mask.
(photo courtesy of J. Auer)

Tethered Diving

THERE ARE MANY CIRCUMSTANCES WHERE DIVERS may need to be directly connected to the surface via a tethered line system. Some examples may include (under certain circumstances) diving in currents, under ice, in zero visibility conditions, and in open ocean/blue water environments. Many divers also use a hard wire communications line that can be incorporated into the tether. This tether can range from a simple line and harness configuration, where divers have a buddy in the water with them, to a single diver configuration described below. Check with your diving safety officer about standards for this type of diving at your institution.

Somers (1986, 1990) describes a compact and portable tethered scuba diving system that has a depth limit of 18 m (60 ft), has a demand breathing mask with communications from diver to surface and vice versa, a primary and emergency breathing gas supply, each with separate first stage regulators, and a surface communications unit. It was designed to be cost effective, easily transportable, have limited required maintenance, and require training of only 16 hours for users who are already scientific scuba certified. This type of tethered diving is useful in very low to zero visibility conditions in shallow water, as a replacement for surface supplied diving. It requires a tender and standby diver as well.

FIGURE 3.6:
Scientific divers may need to be tethered to a line to the surface under certain circumstances, such as in ice diving.
(photo courtesy of J. Heine)

Zero Visibility Conditions

SCIENTIFIC DIVERS SOMETIMES WORK UNDER CONDITIONS where there is no visibility. This can be due to biotic factors (such as an intense plankton bloom), sediment load in the water, or darkness. Entanglement and entrapment are serious problems in zero visibility conditions. Communication by hand signals or slates is impossible, and wireless communications can be very helpful and reassuring to divers in these conditions. Tactile signals, such as squeezes, taps, or line pull signals can also be used.

Training techniques have been developed for zero visibility diving, which can be one of the most stressful activities a diver may encounter (Sellers, 1993). A Zero Visibility Obstacle Course (see also Sellers and Scharf, 1990) has been developed that places emphasis on composure and problem solving abilities in addition to stress. This training uses a blacked-out diver's mask, and a maze consisting of a series of guidelines that connect things like open barrels that the divers must remove their tanks to swim through, gaps in the lines, sharp corners, and potential entanglements. Divers must move slowly and carefully in this environment, using visualization and recognition through tactile senses (Sibthorp, 1995). Scientists desiring to dive in zero-visibility conditions should consult their diving safety officer for training and advice.

Blue Water Diving

DIVING IN OPEN OCEAN or other aquatic environments where the bottom is very deep, is termed blue water diving. Special procedures were developed in the early 1970s to allow for the safe and efficient collection of specimens in this environment (Hamner, 1975). Blue water diving involves perceptions, considerations, and hazards not experienced in many other modes of diving (Heine, 1986; Haddock and Heine, 2005). Without fixed objects for reference, divers can become disoriented. There are also considerations for controlling depth, positioning in the water column, presence of aggressive predators, and the remote nature of many blue water environments.

Blue water diving techniques and equipment consist of a down line to which divers are tethered. A safety diver oversees the entire diving operation. Blue water diving has been conducted for scientific reasons in most of the world's oceans, including polar regions

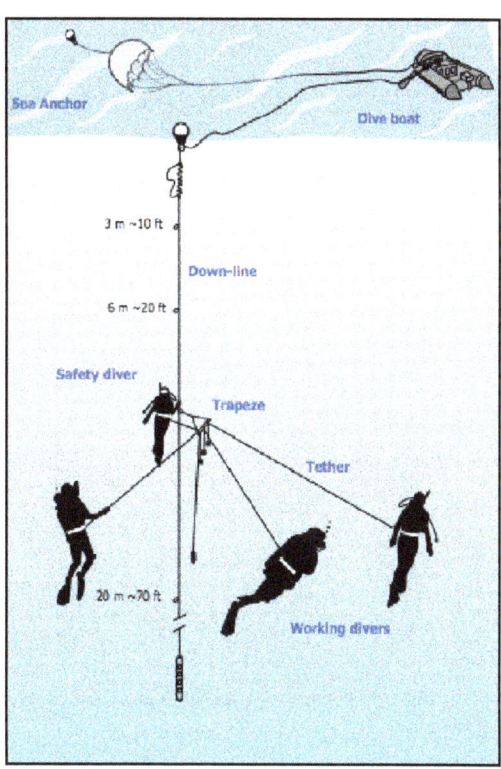

FIGURE 3.7:
Blue water diving rig to control divers in open ocean environment.
(from Haddock and Heine, 2005)

(Sharkey and Griffin, 1987). This specialized mode of diving requires additional training, so see your diving safety officer before attempting to dive in blue water environments. For a complete description of blue water diving, see Haddock and Heine (2005).

Cold Water and Ice Diving

SCIENTIFIC DIVING OCCURS in cold water conditions around the globe. If cold water diving is defined as that which occurs in water temperatures below 4.5°C (40°F), then many freshwater lakes and polar marine environments are included, even in seasons other than winter. These temperatures are sufficiently cold to require special thermal protection and equipment. For example, many single-hose regulators will freeze up and not perform adequately (see regulator section). Dry suits are also commonly used in these cold water environments.

Diving under ice, with the presence of an overhead ceiling, requires a team of surface tenders and special procedures (Heine, 1996; Lang and Sayer, 2007). Considerable training and planning are required before engaging in this diving mode. A special workshop on polar diving by scientists was conducted by the AAUS, and the proceedings may be useful for those planning to dive in polar environments (Lang and Stewart, 1992). A more recent workshop proceedings can be found in Lang and Sayer (2007).

FIGURE 3.8:
Tethered blue water diver estimating density of pelagic organisms.
(photo courtesy of N. Wu)

A considerable amount of scientific diving has been conducted in polar regions and freshwater lakes around the world (Peckham, 1964; Rigler, 1972; Heine, 1996). In Antarctica, scientists began diving in 1957, and continue today. This environment exposes scientists to extreme risk factors such as very cold air and water, winds, a remote location, and relatively deep diving (due to ice scour and anchor ice disturbance in shallow water). Scuba diving is primarily used, but surface-supplied diving has also been conducted in polluted marine environments and the freshwater lakes of Antarctica (Simmons, 1992; Andersen, 2007). Diving in northern polar waters can be quite variable depending upon the season, and is usually conducted in open water from boats (Jewett et al., 2008) or through the ice (Dunton, 1985).

Mixed Gas & Technical Diving

BREATHING GASES OTHER than air can be used to improve safety and increase underwater time. Mixed gas technology is evolving to improve the safety and performance of divers by optimizing a diver's breathing gas during dives, or during various portions of a dive. Gases such as air, nitrox, trimix, heliox, and oxygen are utilized. Nitrox (also known as enriched air, enriched air nitrox, or abbreviated as EAN or EAN_x), is a mixture of lower nitrogen and higher oxygen partial pressures than found in air (Hamilton et al., 1989). Nitrox was introduced into scientific diving by the National Oceanic and Atmospheric Administration (NOAA) in 1978. Two common mixes of nitrox are named NOAA Nitrox I or EAN 32 (a gas mixture consisting of 32% oxygen and 68% nitrogen) and NOAA Nitrox II or EAN 36 (36% oxygen and 64% nitrogen) (Hamilton et al., 1989; Mastro, 1989).

FIGURE 3.9:
Ice diving requires special training and equipment such as tethers, harness, and lights.
(photo courtesy of J. Heine)

The strategy with nitrox is to reduce the nitrogen partial pressure, since nitrogen is the source of decompression and narcosis problems, by replacing it with oxygen, much of which is used metabolically. No-stop times can be greatly increased and decompression times can be reduced using nitrox gas (Miller, 1979). Disadvantages of nitrox include depth limitations, costs, dedicated equipment requirements, and the potential of oxygen toxicity problems.

The National Undersea Research Program (NURP), which is a part of NOAA, supports research on scientific diving techniques. This program is especially interested in wet diving that utilizes nitrox, and it has a program for improving nitrox diving procedures (Smith et al., 1996). The use of nitrox as a breathing gas has steadily increased, as reported to

FIGURE 3.10:
The use of nitrox can extend underwater time for the scientific diver, shown here using a full face mask and video camera.
(photo courtesy of D. Kesling)

the AAUS by Organizational Members, comprising 11.8% of all scientific dives (ca. 14,147 dives) (AAUS, 2011).

For diving beyond the depth limitations of nitrox, other mixes of gases may be appropriate. Trimix gas consists of a mixture of two inert gases such as nitrogen and helium, and oxygen (Bozanic, 2007). Heliox, a mixture of helium and oxygen, is also used by technical and scientific divers (Hamilton, 1990). The individual gas concentrations are predetermined based on the planned depth and specific environmental considerations of the dive. Trimix allows for extended maximum depth and time capabilities, but is expensive, and requires considerable training and equipment. Pure oxygen is sometimes breathed during decompression at shallow depths to hasten the off-gassing of nitrogen.

In addition to alternate breathing gases, divers are often equipped with large and/or multiple cylinders for the additional volume and gas switching requirements of deeper technical diving. Typical configurations might include back-mounted double 120 ft^3 (19 L) cylinders with manifolds, one or more 80 ft^3 (11 L) stage cylinders with decompression gases, and a dry suit inflation cylinder (Bozanic, 2007).

The National Undersea Research Center at the University of North Carolina at Wilmington (NURC), recently conducted dives to 218 fsw on the *Warship Monitor*, using NOAA Trimix 1 tables developed by Hamilton (1993). See Kesling and Shepard (1997) for details. The Caribbean Marine Research Center conducted a workshop to train scientists to dive to depths approaching 300 fsw using various combinations of trimix, nitrox, air, and oxygen (Lombardi, 2003).

Obviously, specific, unique dive tables or computers must be used in accordance with the percentages of nitrogen and helium in such mixtures. Several types of custom dive tables are used by technical divers. Some even use computer programs to plan mixed gas dives. These advanced diving modes make up a relatively small proportion of the scientific dives reported to the AAUS (0.4%, or about 480 mixed-gas dives for the year 2008; AAUS, 2011), and require extensive training (see Bozanic, 2007). The AAUS has standards for mixed gas diving in place (AAUS, 2006).

Rebreathers (Closed Circuit Scuba)

REBREATHERS ARE CLOSED circuit underwater breathing apparatus (ccuba) that recycle a diver's breathing gases. Closed circuit technology can greatly

FIGURE 3.11:
Mixed-gas diver showing equipment configuration with decompression stage cylinder.
(photo courtesy of D. Pence)

extend underwater time based on gas consumption, eliminating the need to carry multiple cylinders. A fully redundant closed circuit rebreather, which recirculates exhaled breathing gas that has been "scrubbed" of carbon dioxide and had oxygen added, could potentially offer a 12-hour gas supply. There can also be a considerable decrease in the expense of gas when compared with open circuit diving (Stanton et al., 2007). This would certainly change the way many scientific divers conduct their research. Potential benefits for scientific diving also include studying animal behavior, and capturing elusive animals such as sea otters, where the absence of bubble noise may be critical to the animals being studied. They might be especially valuable in remote areas where filling scuba cylinders can be a problem (see Bozanic, 2010).

There is a considerable history of rebreather research and development, especially in the military. Hans Hass used a closed circuit oxygen rebreather in the 1930s. NASA used a General Electric closed circuit rebreather in 1964 as part of the Tektite underwater habitat project. Some units use pure oxygen only, and are consequently limited to a depth of 7.6 m (25 ft.) (see Toal, 1989). A semi-closed rebreather emits only a few small bubbles, and consists of a cylinder that can use nitrox gas, breathing bags, absorbent canister, and regulator. Premixed gas flows constantly, providing about two hours in dive duration. Improved thermal balance can be achieved by the heat that is generated from the scrubber canister. These units are relatively simple, without complicated electronics.

Newer, fully automated electronic and computer controlled units are much more sophisticated. For example, the Cis-Lunar MK-5P offers features such as extended depths up to 120 msw (375 fsw), extended underwater times, elimination of nitrogen narcosis if heliox gas is used, gas mix transitions during the dive, noiseless operation, relatively light weight, an incorporated BCD, and a reasonable cost. These rebreathers provide redundant electronic and manual systems for oxygen sensors, depth sensors, microprocessors, and power sources. It even features a through-the-mask electronic heads-up display of oxygen partial pressure, decompression status, and system status.

FIGURE 3.12:
Rebreathers are becoming increasingly available for use in scientific diving.
(photo courtesy of B. Seymour)

Rebreathers have been used in an increasing number of scientific studies. Stanton et al. (2007) report the use of three different types of rebreathers in a forensic study of a shipwreck at a depth of 165 ffw (50 mfw). Expendable supplies were 50% less expensive for the rebreathers as compared to open circuit, and offered decreased fill time and easier logistics. Bozanic (2007) also notes the reduced cost and simplified logistics of using rebreathers in deep lake diving with the National Park Service.

Deep coral reef environments in Puerto Rico have been surveyed using closed-circuit rebreathers (Ambient Pressure Diving Ltd. Inspiration model) (Sherman et al., 2009). Mesophitic coral ecosystems in the depth range of 50-100 m (165-330 ft.) are accessible using trimix gas, with working bottom times of 15 minutes on the deepest dives.

Potential hazards of diving with a rebreather include hyperoxia (which can lead to oxygen toxicity), inhalation of caustic carbon dioxide absorbent material and gases, hypercapnea, hypoxia, DCI, failure of electronics, and contamination from previous users (if not properly disinfected). Vann et al. (2007) report on 80 recreational diving rebreather fatalities from 1998 through 2006, with "equipment troubles" listed as a major triggering event, and "inappropriate gas" listed as a major disabling agent. NOAA has established minimum manufacturing and performance requirements for closed circuit mixed gas rebreathers (Smith and Dinsmore, 2005).

At this time, rebreathers are still relatively expensive, require considerable training and extra maintenance, and are limited in availability. These factors should become less important as more manufacturers and models become available (see Barsky et al., 1998 and

Bozanic, 2010, for a complete review of rebreathers). Approximately 1,079 closed circuit rebreather dives were reported to the AAUS for 2008, and represent 0.9% of the total number of scientific dives performed (AAUS, 2011). The AAUS has standards for rebreather diving in place (AAUS, 2006).

Underwater Habitats and Saturation Diving

AS ALL DIVERS SHOULD KNOW, DECOMPRESSION illness does not occur while the diver is underwater at depth. The problems only arise upon surfacing, where the ambient pressure is reduced. In theory, a person could stay underwater indefinitely and not suffer from DCI. People have conducted extended underwater missions, on the order of weeks underwater, in what is called saturation diving. Saturation means that the tissues are saturated with nitrogen. Saturation diving is done in an underwater habitat placed on the seafloor and pressurized near the ambient pressure surrounding the habitat. This allows the diving scientists to conduct extended dives (termed excursions) without the decompression obligations that would be incurred if they were returning to the surface following the dive. It is not uncommon to have dive times of 6 hours per day, for many consecutive days. It has been estimated that saturation missions of 8-10 days would take at least 40-60 days if conducted using surface-based technology (Shepard et al., 1996).

A number of different breathing gas mixtures can be used for saturation diving, based on the depths involved. Gases such as air, nitrox, heliox, and helium-hydrogen-oxygen mixtures have been used. Depending on the breathing gas, depth, and times of saturation, decompression times may be on the order of 16 to 64 hours.

FIGURE 3.13:
Saturation diving is conducted in the Aquarius habitat which is operated by NOAA.
(photo courtesy of M. Hay)

The National Oceanic and Atmospheric Administration (NOAA) currently operates the Aquarius Undersea Laboratory in the Florida Keys, through the National Undersea Research Program (NURP) at the University of North Carolina at Wilmington. This habitat is a major advance in technology, providing research space and a comfortable living environment for up to five scientists and one technician. A four-day training program is required for visiting scientific divers to familiarize them with operational and safety procedures (Kesling and Berey, 1989; http://aquarius.uncw.edu/).

Lock-out Diving

SOME SUBMERSIBLES HAVE THE CAPABILITY to allow a diver to "lock-out." The diver is transported to the dive site inside the submersible, and exits the submersible, usually by means of umbilical-supplied diving from the submersible. At the completion of the dive, the diver enters the submersible for transportation back to the surface. This type of diving is useful for sampling habitats or organisms that are inaccessible from the surface or by ROV or submersible manipulators (Flemming and Max, 1996).

FIGURE 3.14:
Lock-out diver and the Johnson-Sea-Link research submersible.
(photo courtesy of Harbor Branch Oceanographic Institution)

Diving Bell

AN OPEN WATER DIVING BELL IS USED in the commercial diving industry for extended diving times in deeper water. These diving bells function as a work platform that is lowered to depth from the surface. They can be raised and lowered with the divers depending on the tasks and decompression requirements. While they are rarely used for scientific diving, they may play an important role in the future, as scientific diving extends its range and modes to explore new underwater environments.

Deep-diving systems, utilizing a submersible-diving chamber that is lowered to the working depth, have been used by scientific divers for pelagic and benthic research to depths of 260 m (860 ft), in conjunction with the U. S. Navy (Orzech and Nealson, 1985). They are commonly used in the commercial diving field.

Diver Propulsion Vehicles and Towed Sleds

WHEN DIVERS WANT TO COVER A LARGE AREA underwater for initial evaluations or survey purposes, they may use a diver propulsion vehicle (DPV) or a sled towed from behind a boat (Sigl, 1969; Flemming and Max, 1996). This allows the diver to use propulsion other than their legs, extending the distance that can be covered on a dive and lowering air consumption. The DPV is a battery-powered propulsion device that uses a propeller for thrust. The diver can generally control the speed and direction of the DPV by adjusting the trigger and angle of the unit. Speeds of up to 5.2 kph (3.2 mph) and running times up to 90 minutes can be achieved, yielding an effective range of up to 4,550 m (15,000 ft). Some DPVs have places to mount a depth gauge and compass, and also a video mount for surveying purposes.

Towing a diver behind a boat may be practical if the water is relatively clear and free of obstructions such as kelp, vertical walls, etc. A planing board is towed about 25 m (80 ft) behind the boat, and the diver holds onto two handles (Dart and Rainbow, 1976; Lewis, 1987). Some models of tow boards hold two divers (Lewis, 1987; Swan, 1987). The speed of the boat and the depth of the diver must be closely monitored. Since the diver cannot record data on a slate, communications to the surface are essential. This may also be required since a single diver does not have a buddy in the water. Many hundreds of meters can be covered on a single dive.

The Pacific Islands Fisheries Science Center of the National Marine Fisheries Service uses towboarding with snorkelers or scuba divers (Boland, 2005). The board is towed from 50 to 200 feet behind a small vessel. The diver can control ascent, descent, and left or right steering. This technique has been used to survey marine debris, coral reefs, and large fish to a depth of 120 feet. The board can be configured to carry digital imaging devices and even a small CTD.

Vroom et al. (2010) collected qualitative and quantitative benthic composition data at a level of functional group using towed-diver surveys in Hawaii. They towed two scuba divers 60 m behind a small boat, at a speed of around 1.5 knots. Diver positions were recorded with a GPS receiver on the vessel. A 50 minute survey was conducted, and was broken down into five-minute segments to estimate categories such as live and stressed coral, crustose coralline algae, macroalgae, and substrate (see also Kenyon et al., 2006).

The towed-diver survey methodology has been used to study large reef fishes that occur in relatively low densities where encounter rates are low (Richards et al., 2011). The neutral buoyancy towboard was constructed of plywood and outfitted with a digital video camera, temperature/depth recorder, timers, and data sheets. Communications with the surface vessel were performed using a magnetic switch telegraph device.

There are some safety considerations that must be adhered to when utilizing this technique. There are the possibilities of rapid ascent, entanglement in lines, ear clearing, flooded masks, regulator loss, and separation from the board. For a detailed description of the technique, see Boland (2005) and Richards et al. (2011).

FIGURE 3.15:
Scientific divers sometimes use diver propulsion vehicles to evaluate large areas quickly.
(photo courtesy of K. Flanagan)

FIGURE 3.16 :
Scientific diver using a towboard for mesoscale fishery-independent survey assessment.
(photo courtesy of Bulletin of Marine Science, 87(1): 55-74)

FIGURE 3.17:
Diving around underwater vehicles such as ROVs has unique hazards.
(photo courtesy of B. Seymour)

FIGURE 3.18:
Diving around vehicles such as the Autonomous Benthic Explorer (ABE) requires special training.
(photo courtesy of T. Rioux)

FIGURE 3.19:
Scientific divers use lift bags to move instruments and other experimental equipment underwater.
(photo courtesy of G. Eyal)

A specialized cinematography sled has been used in applications such as towed fishing gear evaluation, surveys, and search and recovery operations (Workman and Watson, 1986; Swan, 1987; Richards et al., 2011).

Diving With ROVs

REMOTELY OPERATED vehicles (ROVs) are used for a variety of scientific operations. These unmanned vehicles operate via an umbilical to the surface that provides power and transmits signals and data. If divers and ROVs are working together in the water, special care must be taken to provide for a safe environment for the diver. Special considerations should include the potential for electric shock, entanglement, collision danger, injury from mechanical thrusters, lasers, and high frequency output from sonar (Caramanna, 2005; C. J. Smith, pers. comm.). Standards for commercial diving with ROV's have been established by the Association of Diving Contractors (ADC, 1994).

Lift Bags

A SCIENTIFIC DIVER MAY need to lift or lower relatively heavy objects such as bottom weights, cages, rocks, or instruments. The diver's buoyancy compensator should never be used as a lifting device because of the potential of an uncontrolled rapid ascent if the item lifted is dropped. As a general rule, anything that weighs more than about 5 kg (ca. 10 lb) should be moved with a lift bag. Special training is required in their use, and large projects should be left to commercial divers.

There are a number of types and sizes of lift bags available, each with specific applications. Open bottom lift bags look like a small parachute, and come in sizes of lifting capacities from 23 to 5,500 kg (ca. 50 to 12,000 lb). These bags are versatile, and have diver-controlled dump valves to vent off expanding air on ascent. Pillow bags, boulder bags, and ocean salvage tubes are totally enclosed, and are commonly used for raising sunken vessels or for surface towing of heavy objects.

Cave and Cavern Diving

SCIENTIFIC CAVE DIVING is a relatively new endeavor, practiced by a few well-trained individuals (Skiles, 1986). Due to its recent application in scientific diving, the level of research centers on general mapping and surveys, biological collections and taxonomy, distribution and abundance of organisms, identification of archaeological sites, and basic hydrological studies (Bozanic, 1997). This type of diving is very equipment-dependent, and requires extensive training, planning, and resources. It is not unusual for cave divers to practice multiple staging of cylinders along the route, to carry multiple cylinders (and even different types of breathing gases) during a dive, and to have extended decompression times at the end of long and/or deep dives. They also use special reels, multiple redundant lights, extra long regulator hoses, and special techniques for kicking and buoyancy control to minimize the potential for stirring up sediment in the water.

FIGURE 3.20:
Cave divers use a variety of redundant equipment and special techniques.
(photo courtesy of J. Heine)

Kernagis et al. (2008) describe the detailed logistics required to accomplish an extreme cave traverse. Using a standardized approach of pre-dive planning, gas and equipment preparation, review of the accident management plan, assembling support logistics, and conducting the dive, they were able to traverse 11.2 km (7 mi.) at a maximum depth of 95 mfw (300 ffw), for a total underwater time of over 21 hours.

The AAUS has standards for scientific cave and cavern diving in place (AAUS, 2006).

FIGURE 3.21:
Sket bottles incorporate one-way valves, and may be used to collect multiple animals in a single bottle.
(photo courtesy of J. Bozanic)

Night Diving

SCIENTISTS MAY NEED TO DIVE AT NIGHT TO COLLECT certain organisms, make behavioral observations, or collect samples. This obviously requires the use of lights for safety and logistical reasons. There are a variety of hand-held lights available in varying intensities and beam angles, and a choice of bulbs or lamps such as halogen or LED. High-intensity lights used mainly in cave diving are 10 watt HID canister lights with rechargeable NiMH or Lithium-ion batteries. There are head-mounted lighting systems available that allow the diver's hands to be free for other tasks (e.g., NiteRider).

Training Exercises and Lead Diver

FOR THE TRAINING EXERCISES LISTED THROUGHOUT this book, it is suggested that students in a scientific diving training program should rotate through the "lead diver" position, so that everyone has an opportunity to experience the responsibilities associated with that position. The AAUS describes the lead diver as follows:

For each dive, one individual shall be designated as the lead diver. He/she shall be at the dive location during the diving operation. The lead diver shall be responsible for the following:

Coordination with other known activities in the vicinity that are likely to interfere with diving operations.

Ensuring all dive team members possess current certification and are qualified for the type of diving operation.

Planning dives in accordance with program requirements.

Ensuring safety and emergency equipment is in working order and at the dive site.

Briefing the dive team members on:

FIGURE 3.22:
Pool training can be especially valuable for learning techniques to be used in the field.
(photo courtesy of J. Auer)

Dive objectives.

> Unusual hazards or environmental conditions likely to affect the safety of the diving operation.
>
> Modifications to diving or emergency procedures necessitated by the specific diving operation.
>
> Suspending diving operations if in his/her opinion conditions are not safe.
>
> Reporting to the DSO and Diving Control Board any physical problems or adverse physiological effects, including symptoms of pressure-related injuries.

These training exercises can often be included as a part of an actual scientific research project, utilizing the labor of a scientific diving class for an actual project. It is often beneficial to have a dry run of each exercise on land before the dives. This has proven to save time and improve the diver's understanding of the objectives of the dives.

TRAINING EXERCISE #1
Introduction to Specialized Equipment and Techniques

Equipment and/or environment needed: full face masks, surface supplied diving equipment, hookah, communications, tethered diving, zero visibility, blue water diving, cold water or ice diving, nitrox, mixed gas, rebreathers, diver propulsion vehicle, towed sled, lift bags, night diving (as available or desired).

Objective: To briefly introduce scientific divers to a variety of specialized equipment or techniques unique to the diving program or area.

This training exercise is designed to briefly expose scientific divers to the variety of specialized equipment and techniques that are available in the program. Equipment and expertise can often be borrowed from local institutions. For most of the exercises, classroom lectures should be given prior to the dives. A thorough briefing must be given by the teaching staff at the site prior to the dive. A "dry run" on land is often beneficial prior to the dive.

A full day of diving may be required for each topic or mode, depending upon the size of the class and the number of staff members available. It is important to explain the complexity of using some of this equipment or modes of diving, including the preparation time and post-dive maintenance that may be required. Finally, it must be stressed that these are not "certification" dives, but merely introductions to the methods. The diving safety officer will be able to explain what exactly is required for authorization to use certain equipment or modes of diving.

Equipment Manufacturers and Suppliers

Brownie's Third Lung (hookah, rebreathers, regulators)
940 NW 1st St.
Ft. Lauderdale, FL 33311
(954) 462-5570
www.browniedive.com

Carter Lift Bag, Inc.
29500 Green River Gorge Road
Enumclaw, WA 98022
(800) 4-LIFTBAG
www.carterbag.com

Cochran Undersea Technology (rebreathers, dive computers)
1758 Firman Dr.
Richardson, TX 75081
(972) 644-6284
www.divecochran.com

DEMA: Diving Equipment & Marketing Association
3750 Convoy St., Ste. 310
San Diego, CA 92111-3741
(858) 616-6408
www.dema.org

Dive Rite (technical scuba equipment)
175 NW Washington St.
Lake City, FL 32055
(386) 752-1087
www.diverite.com

Diving Unlimited International (DUI) (drysuits)
1148 Delevan Dr.
San Diego, CA 92102-2499
(800) 325-8439
www.dui-online.com

Draeger Safety, Inc. (rebreathers)
101 Technology Drive
Pittsburgh, PA 15275
(412) 787-8383
www.draeger.com

Efcom. Inc. (pingers, communications)
1336 E. Wilshire Blvd.
Santa Ana, CA 92705
(714) 543-0677

Farallon USA (diver propulsion vehicles)
World Exchange Plaza
45 O'Connor St., Ste. 1150
Ottawa, Ontario
Canada
K1P 1A4
(613) 755-4048
www.farallonusa.com

Interspiro (AGA full face mask)
10225 82nd Ave.
Pleasant Prairie, WI 53158-5801
(800) 468-7788
www.interspiro.com

Kirby Morgan Dive Systems (surface supplied diving, full face masks)
1430 Jason Way
Santa Maria, CA 93455
(805) 928-7772
www.kmdsi.com

Ocean Technology Systems (underwater communications; Buddy Phone)
3133 West Harvard St.
Santa Ana, CA 92704
(714) 754-7848
www.oceantechnologysystems.com

Steam Machines, Inc. (rebreathers)
538 Hiwassee Rd.
Lebanon, TN 37087-9298
(615) 374-0202
www.steammachines.com

Torpedo, Inc. (diver propulsion vehicles)
1334 Spalding Rd.
Dunedin, FL 34698
(800) 489-6774
www.torpedodpv.com

Trelleborg Viking (drysuits)
290 Forbes Blvd.
Mansfield, MA 02048
(774) 719-1444
www.vikingdiving.com

Continuing Education and Training

American Academy of Underwater Sciences (AAUS)
Dauphin Island Sea Lab
101 Bienville Blvd.
Dauphin Island, AL 36528
(251) 591-3775
Email: aaus@disl.org
www.aaus.org

American Nitrox Divers International (ANDI)
74 Woodcleft Ave.
Freeport, NY 11520
(516) 546-2026
www.andihq.com

Divers Alert Network (DAN) (oxygen training, insurance)
6 West Colony Place
Durham, NC 27705
(919) 684-2948
www.diversalertnetwork.org

International Association of Nitrox and Technical Divers (IANTD)
2124 NE 123rd St., Ste. 210
North Miami, FL 33181-2939
Miami Shores, FL 33138
(786) 704-9722
www.iantd.com

National Association for Cave Diving
P.O. Box 14492
Gainesville, FL 32604
(386) 497-3011
www.safecavediving.com

National Oceanic and Atmospheric Administration (NOAA)
National Undersea Research Program (NURP)
1335 East-West Highway
Silver Spring, MD 20910
(301) 713-2427
www.ucc.uconn.edu/~wwwnurc/nurp.html

National Speleological Society Cave Diving Section
295 NW Commons Loop
Suite 115-317
Lake City, FL 32055
(386) 454-5550
www.nsscds.org/test/drupal/index.php

NAUI Worldwide
P.O. Box 89789
Tampa, FL 33689-0413
(813) 628-6284
www.naui.org

PADI
30151 Tomas
Rancho Santa Margarita, CA 92688-2125
(949) 858-7234
www.padi.com

Technical Diving International
18 Elm St.
Topsham, ME 04086
(207) 729-4201
www.tdisdi.com

References

American Academy of Underwater Sciences. 1996. Standards for Scientific Diving Certification and Operation of Scientific Diving Programs. 54 pp.

American Academy of Underwater Sciences. 2006. Standards for Scientific Diving. 80 pp.

Andersen, D.T. 2007. Antarctic inland waters: Scientific diving in the perennially ice-covered lakes of the McMurdo Dry Valleys and Bunger Hills. In: Lang, M.A. and M.D.J. Sayer (eds.). 2007. Proc. Intl. Polar Diving Workshop. Smithsonian Institution, Wash. D.C. pp. 163-169.

Association of Diving Contractors, Inc. 1994. Consensus Standards for Commercial Diving Operations. ADC, Houston, TX.

Bachrach, A.J. 1985. Cold stress and the scientific diver. In: Mitchell, C.T. (ed.), Diving for Science...85, Proceedings of Joint International Scientific Diving Symposium, Amer. Acad. Underwater Sci. pp. 31-37.

Barsky, S.M. 1986. Technical specifications of a system for contaminated water diving. In: Mitchell, C.T. (ed.), Diving for Science...86, Proceedings of Amer. Acad. Underwater Sci. pp. 27-32.

Barsky, S. 1990. Diving in high risk environments. Dive Rescue, Inc., Fort Collins, CO. 118 pp.

Barsky, S. Diving with the Divator MK II full face mask. Avail. from Best Publishing.

Barsky, S. Diving with the EXO-26 full face mask. Avail. from Best Publishing.

Barsky, S., D. Long, and B. Stinton. 1992. Dry Suit Diving: A Guide to Diving Dry. Watersport Publishing, San Diego, CA. 185 pp.

Barsky, S., M. Thurlow, and M. Ward. 1998. The simple guide to rebreather diving. Best Publishing, Flagstaff, AZ.

Barsky, S. 2010. Choosing and using full-face masks. DVD from Hammerhead Press, Ventura, CA.

Boland, R. 2005. The Pacific Islands Fisheries Science Center dive program: Meeting the challenges of the Pacific region. In: Godfrey, J.M. and S.E. Shumway, (eds.), Diving for Science 2005, Proceedings of Amer. Acad. Underwater Sci. pp. 179-189.

Bozanic, J. and J. Mastro. 1992. Regulator function in the Antarctic. In: M.A. Lang and J.R. Stewart (eds.), Proceedings of Amer. Acad. Underwater Sci. Polar Diving Workshop. pp. 18-27.

Bozanic, J.E. 1997. AAUS standards for scientific diving operations in cave and cavern environments: A proposal. In: Maney, Jr. E.J. and C.H. Ellis, Jr. (eds.), Diving for Science...1997, Proceedings of Amer. Acad. Underwater Sci. pp. 5-24.

Bozanic, J.E. 2002. Mastering Rebreathers. Best Publishing, Flagstaff, AZ. 548 pp.

Bozanic, J.E. 2007. An evolution of scientific mixed gas diving procedures at the National Park Services Submerged Resources Center. In: Pollock, N.W. and J.M. Godfrey (eds.), Diving for Science 2007, Proceedings of Amer. Acad. Underwater Sci. pp. 143-154.

Bozanic, J.E. 2010. Mastering Rebreathers, second ed., Best Publishing, Flagstaff, AZ. 704 pp.

Caramanna, G. 2005. Scientific diving and ROV techniques applied to the geomorphological and hydrogeological study of the world's deepest karst sinkhole, (Pozzo del Merro-Latium-Italy). In: Godfrey, J.M. and S.E. Shumway, (eds.), Diving for Science 2005, Proceedings of Amer. Acad. Underwater Sci. pp. 233-238.

Clarke, J.R. 2007. Scuba regulators for use in cold water: The U.S. Navy perspective. In: Lang, M.A. and M.D.J. Sayer (eds.). Proc. Intl. Polar Diving Workshop. Smithsonian Institution, Wash. D.C. pp. 35-44.

Colwell, R.R. (ed.). 1982. Microbial hazards of diving in polluted waters. Maryland Sea Grant Pub. No. UM-SG-TS-82-01.

Dart, J. K.G. and P.S. Rainbow. 1976. Estimating echinoderm populations. In: Drew, E.A., J.N. Lythgoe, and J.D. Woods (eds.). 1976. Underwater research. Academic Press, New York. 430 pp.

DeWreede, R.E. 1985. Destructive (harvest) sampling, pp. 147-160. In: Littler, M.M. and D.S. Littler (eds.). Handbook of Phycological Methods. Ecological Field Methods: Macroalgae. Cambridge University Press. 617 pp.

Dowling, G. 1963. Diver's instrumented observation board. U.S. Navy Mine Defense Laboratory Report 210 (Arlington, VA, U.S. Navy Government Clearinghouse).

Drew, E.A., J.N. Lythgoe, and J.D. Woods (eds.). 1976. Underwater research. Academic Press, New York. 430 pp.

Dunton, K.H. 1985. Growth of dark-exposed *Laminaria saccharina* (L) Lamour and *Laminaria solidungula* J Ag (Laminariales, Phaeophyta) in the Alaskan Beaufort Sea. J. Exp. Mar. Biol. And Ecol. 94: 181-189.

Flemming, N.C. and M.D. Max. 1996. Scientific Diving: A General Code of Practice. Best Publishing, 278 pp.

Gilliam, B. and R. von Maier. Deep Diving: An Advanced Guide to Physiology, Procedures, and Systems. Watersport Publishing. 1993.

Haddock, S.H.D. and J.N. Heine. 2005. Scientific Blue-Water Diving Guidelines. Calif. Sea Grant Publ. No. T-057. Univ. of Calif., La Jolla, CA 92093. 49 pp.

Hamilton, R.W. 1990. The warm mineral springs decompression plan and tables. In: W.C. Jaap, (ed.), Diving for Science...1990, Proceedings of Amer. Acad. Underwater Sci. pp. 147-156.

Hamilton, R.W. 1993. NOAA/HRL trimix decompression tables for dives on the USS Monitor. Hamilton Research, Ltd. NY.

Hamilton, R.W., D.J. Crosson, and A.W. Hulbert (eds.). 1989. Workshop on enriched air nitrox diving. National Undersea Research Program Report 89-1. 153 pp.

Hamner, W.M. 1975. Underwater observations of blue-water plankton: Logistics, techniques, and safety procedures for divers at sea. Limnol. Oceanogr. 20(6): 1045-1051.

Heine, J. N. (ed.). 1986. Blue water diving guidelines. Calif. Sea Grant Publ. no. T-CS-GCP-014. 46 pp.

Heine, J.N. 1996. Cold Water Diving: A Guide to Ice Diving. Best Publishing. 127 pp.

Jewett, S.C., R. Brewer, H. Chenelot, R. Clark, D. Dasher, S. Harper, and M. Hoberg. 2008. Scuba techniques for the Alaska Monitoring and Assessment Program (AKMAP) of the Aleutian Islands, Alaska. In: Brueggeman P. and N.W. Pollock (eds.). Diving for Science 2008, Proceedings of the American Academy of Underwater Sciences, pp. 71-89.

Kenyon, J.C., R.E. Brainard, R.K. Hoeke, et al. 2006. Towed-diver surveys, a method for mesoscale spatial assessment of benthic reef habitat: a case study at Midway Atoll in the Hawaiian Archipelago. Coast Manag. 34: 339-349.

Kesling, D. and R. Berey. 1989. Training, equipment, and operational procedures for conducting scientific saturation diving activities. In: Lang, M.A. and W.C. Jaap, (eds.), Diving for Science...1989, Proceedings of Amer. Acad. Underwater Sci. pp. 199-211.

Kesling, D.E. and A.N. Shepard. 1997. Scientific diving beyond 190 FSW! The NURC/UNCW experience. In: Maney, Jr. E.J. and C.H. Ellis, Jr. (eds.), Diving for Science...1997, Proceedings of Amer. Acad. Underwater Sci. pp. 101-104.

Lang, M.A. and G.H. Egstrom (eds.). 1990. Biomechanics of Safe Ascents Workshop. Proceedings of the American Academy of Underwater Sciences. 220 pp.

Lang, M.A. and R.D. Vann (eds.). 1992. Repetitive Diving. Proceedings of the American Academy of Underwater Sciences. 339 pp.

Lang, M.A. and J.R. Stewart (eds.). 1992. Polar Diving Workshop. Proceedings of the American Academy of Underwater Sciences. 100 pp.

Lang, M.A. and M.D.J. Sayer (eds.). 2007. Proc. Intl. Polar Diving Workshop. Smithsonian Institution, Wash. D.C. 213 pp.

Lang, M.A. and A.O. Brubakk (eds.). 2009. The future of diving: 100 years of Haldane and beyond. Smithsonian Institution Scholarly Press. 286 pp.

Lewis, C. 1987. Coldwater SCUBA diving search and recovery operations. In: M.A. Lang (ed.), Coldwater Diving for Science...1987, Proceedings of the American Academy of Underwater Sciences. pp. 151-161.

Lombardi, M. 2003. Considerations for scientific technical diving: An overview of logistics, procedures, and implications for program development. In: S.F. Norton (ed.), Diving for Science 2003, Proceedings of Amer. Acad. Underwater Sci. pp. 43-58.

Marelli, D.C. and W. Jaap. 1990. Description of a low-cost, shallow-water, surface-supplied diving system. In: W.C. Jaap, (ed.), Diving for Science...1990, Proceedings of Amer. Acad. Underwater Sci. pp. 267-271.

Mastro, S.J. 1989. Use of two primary breathing mixtures for enriched air diving operations. In: M.A. Lang (ed.), Advances in Underwater Science...1988, Proceedings of Amer. Acad. Underwater Sci. pp. 241-247.

Mastro, J.G. and N.W. Pollock. 1995. Sherwood maximus regulator temperature and performance during Antarctic diving. In: Harper, D.E. (ed.), Diving for Science...1995, Proceedings of Amer. Acad. Underwater Sci. pp. 53-62.

Miller, J.W. (ed.). 1979. NOAA Diving Manual: Diving for Science and Technology Second edition, National Oceanic and Atmospheric Administration, U.S. Dept. of Commerce.

Mount, T. Technical EAN_x. International Association of Nitrox and Technical Divers.

Mount, T. Trimix. International Association of Nitrox and Technical Divers.

Mount, T. Cave Diving. International Association of Nitrox and Technical Divers.

Mount, T. and B. Gilliam. Mixed Gas Diving: The Ultimate Challenge for Technical Diving. Watersport Publishing, 1993.

Murdoch, T.E. 1987. Optimizing acoustic communications for quality and range in diving and submersible applications. In: M.A. Lang (ed.), Coldwater Diving for Science...1987, Proceedings of the American Academy of Underwater Sciences. pp. 191-203.

Orzech, J.K. and K.H. Nealson. 1985. The use of deep-diving systems in marine research. In: Mitchell, C.T. (ed.), Diving for Science...85, Proceedings of Joint International Scientific Diving Symposium, Amer. Acad. Underwater Sci. pp. 31-37.

Palmer, R. An Introduction to Technical Diving. Underwater World Publications Ltd. 1994. Avail. from Best Publishing.

Paulet, S.B and D.L. McLean. 1989. Diver efficiency: Low drag, maximum thrust, and the check is in the mail. In: M.A. Lang (ed.), Advances in Underwater Science...1988, Proceedings of Amer. Acad. Underwater Sci. pp. 249-252.

Peckham, V. 1964. Year-round scuba diving in the Antarctic. Polar Record 12(77): 143-146.

Phoel, W.C. 1981. NOAA's requirements and capabilities for diving in polluted waters, In: Microbial Hazards of Diving in Polluted Waters, Mar. Tech. Soc. Journal, Vol. 15, No. 2, pp. 4-10.

Richards, B.L., I.D. Williams, M.O. Nadon, and B.J. Zgliczynski. 2011. A towed-diver survey method for mesoscale fishery-independent assessment of large-bodied reef fishes. Bull. Mar. Sci. 87(1): 55-74.

Rigler, F.H. 1972. Director's review. In: Char Lake Project Annual Report 1971-72. Canadian Committee International Biological Program.

Ronnestad, I., H.P. Roverud, I. Strand, A. Hope, and R. Vaernes. 1987. Thermal and mental effects of wet and dry suit dives to 18 meters. In: M.A. Lang (ed.), Coldwater Diving for Science...1987, Proceedings of the American Academy of Underwater Sciences. pp. 237-244.

Sellers, S.H. and R.J. Scharf. 1990. Training scientific divers for zero visibility diving. In: W.C. Jaap, (ed.), Diving for Science...1990, Proceedings of the American Academy of Underwater Sciences.

Sellers, S.H. 1993. Bridging the experience gap: Techniques for reducing the stress of zero visibility training. In: Heine, J.N. and N.L. Crane, (eds.), Diving for Science...1993, Proceedings of Amer. Acad. Underwater Sci. pp. 127-131.

Sharkey, P.I. and J.J. Griffin. 1987. Blue-water diving north of the Arctic circle. In: M.A. Lang (ed.), Coldwater Diving for Science...1987, Proceedings of the American Academy of Underwater Sciences. pp. 245-256.

Sheldrake, S., D. Davoli, M. Poulsen, P.B. Duncan, and R. Pederson. 2009. Diver exposure scenario for the Portland Harbor risk assessment. In: Pollock, N.W. (ed.), Diving for Science 2009, Proceedings of Amer. Acad. Underwater Sci., Sea Grant Publ. No. CTSG-10-09. pp. 7-18.

Shepard, A.N., D.A. Dinsmore, S.L. Miller, C.B. Cooper, and R.I. Wicklund. 1996. Aquarius Undersea Laboratory: The next generation. In: Lang, M.A. and C.C. Baldwin (eds.), Methods and Techniques of Underwater Science, Proceedings of Amer. Acad. Underwater Sci. pp. 205-212.

Sherman, C., R. Appeldoorn, M. Carlo, M. Nemeth, H. Ruiz, and I. Bejarano. 2009. Use of technical diving to study deep reef environments in Puerto Rico. In: Pollock, N.W. (ed.), Diving for Science 2009, Proceedings of Amer. Acad. Underwater Sci., Sea Grant Publ. No. CTSG-10-09. pp. 58-65.

Sibthorp, R.J. 1995. Full-face-mask and zero-visibility training for scientific blackwater divers. In: Harper, D.E. (ed.), Diving for Science...1995, Proceedings of Amer. Acad. Underwater Sci. pp. 95-102.

Sigl, W., V. VonRad, H.J. Oeltzschner, K. Braune, and F. Fabricius. 1969. Diving sled: a tool to increase the efficiency of underwater mapping by scuba divers. Mar. Geol. 7:357-363.

Simmons, G.M. 1992. Surface-supplied diving in freshwater lakes. In: M.A. Lang and J.R. Stewart (eds.), Proceedings of Amer. Acad. Underwater Sci. Polar Diving Workshop. pp. 28-31.

Skiles, W.C. 1986. The scientific future of cave diving. In: Mitchell, C.T. (ed.), Diving for Science...86, Proceedings of Amer. Acad. Underwater Sci. pp. 37-46.

Smith, N.E., B.S.P. Moore, M. Collie, A. Kalvaitis, E.P. Myers, and J.M. Zeidner. 1996. NOAA's national undersea research program: 1996/1997 overview. In: Lang, M.A. and C.C. Baldwin (eds.), Methods and Techniques of Underwater Science, Proceedings of Amer. Acad. Underwater Sci. pp. 213-219.

Smith, N.E. and D.A. Dinsmore. 2005. Evolution of the NOAA minimum manufacturing and performance requirements for closed circuit mixed gas rebreathers. In: Godfrey, J.M. and S.E. Shumway (eds.), Diving for Science 2005, Proceedings of Amer. Acad. Underwater Sci. pp. 43-48.

Somers, L.H. 1986. A compact and portable diving system for scientists. In: Mitchell, C.T. (ed.), Diving for Science...86, Proceedings of Amer. Acad. Underwater Sci. pp. 13-25.

Somers, L.H. 1987. A portable diving system for search & rescue, scientific, and commercial divers. Michigan Sea Grant Program, Ann Arbor, MI. 21 pp.

Somers, L.H. 1990. University of Michigan Diving Manual, Vol. II. Underwater Research Methods. 182 pp.

Somers, L.H. 1990. Use of tethered scuba diving to improve safety and efficiency. In: W.C. Jaap, (ed.), Diving for Science...1990, Proceedings of Amer. Acad. Underwater Sci. pp. 345-356.

Stanton, G., K. Meverden, T. Thomsen and J. Garey. 2007. Closed-circuit rebreathers in the forensic study of the *Rouse Simmons* shipwreck. In: Pollock, N.W. and J.M. Godfrey (eds.), Diving for Science 2007, Proceedings of Amer. Acad. Underwater Sci. pp. 89-99.

Somers, L. Technical Diving. Unpubl. Manuscript.

Stanton, G., W. Burnett, and B. Gilam. 1994. Polluted water technology used in a miramictic lake in Belau, Micronesia. In: Petrecca, R. (ed.), Diving for Science, 1994, Proceedings of Amer. Acad. Underwater Sci. pp. 81-91.

Stinton, R.T. 2007. A review of diver thermal protection strategies for polar diving: present and future. In: Lang, M.A. and M.D.J. Sayer (eds.). 2007. Proc. Intl. Polar Diving Workshop. Smithsonian Institution, Wash. D.C. pp. 13-34.

Swan, G.A. 1987. Use of a diver's sled and laser location system for salmon spawning surveys. In: M.A. Lang (ed.), Coldwater Diving for Science...1987, Proceedings of the American Academy of Underwater Sciences. pp. 265-277.

Toal, F.J. 1989. Closed circuit oxygen rebreathers - a new approach. In: M.A. Lang (ed.), Advances in Underwater Science...1988, Proceedings of Amer. Acad. Underwater Sci. pp. 321-324.

Vann, R.D., N.W. Pollock and P.J. Denoble. 2007. Rebreather fatality investigation. In: Pollock, N.W. and J.M. Godfrey (eds.), Diving for Science 2007, Proceedings of Amer. Acad. Underwater Sci. pp. 101-110.

Vroom, P.S., C.A. Musburger, S.W. Cooper, J.E. Maragos, K.N. Page-Albins, and M.A.V. Timmers. 2010. Marine biological community baselines in unimpacted tropical ecosystems: spatial and temporal analysis of reefs at Howland and Baker Islands. Biodivers. Conserv. 19: 797-812.

Walker, C.H. and J.J. Gurney. 1985. The recovery of diamonds from the surf zone of the South Atlantic near the Olifants River, R.S.A. In: Mitchell, C.T. (ed.), Diving for Science...85, Proceedings of Joint International Scientific Diving Symposium, Amer. Acad. Underwater Sci. pp. 318-330.

Workman, I.K. and J.W. Watson. 1986. Construction and operation of a two place diver's sled. Unpubl. ms., 19 pp., NOAA/NMFS, Pascagoula Facility, P.O. Drawer 1207, Pascagoula, MS 39567.

Youngbluth, M.J. 1983. Manned submersibles and sophisticated instrumentation: Tools for oceanographic research. Subtech 83, paper 7.1, 6 pp.

CHAPTER 4

LOCATING, MARKING, AND MAPPING SITES

LOCATING, RELOCATING, AND ADEQUATELY MARKING a study site is critical, especially in areas of low water visibility. On the surface, there are a number of different methods that can be employed. It is a good idea to use a couple of different methods, in order to have a back up if one method fails. A relatively crude method is to use compass bearings towards large, readily identifiable objects on land. For much higher accuracy (ca. +/- 5 m), these bearings can be used in conjunction with shore lineups, where pairs of objects that are in a straight line can be used to triangulate a position. Of course, the major disadvantage of these methods is that the shore markers may not be visible during bad weather (fog, rain, etc.) or at night.

Modern Global Positioning System (GPS) units are now relatively inexpensive, portable, and accurate within about 15 m (50 ft). Units with differential capabilities are accurate within about 3 to 5 m (10 to 15 ft). They also have the advantage of being able to store many locations (waypoints), give the heading, distance, and time to each waypoint from your present position, and store multiple routes with many legs on each route. Many units also have graphical navigation screens. GPS has essentially replaced LORAN in most areas of the world. At this time, GPS cannot be used underwater without a surface unit, but this may change soon (Coudeville and Thomas, 1998; ACSA). Radar can also be used if available.

A rapid and easy method of determining the depth of the water is by using a hand-held sonar device. A unit the size of a small flashlight can instantly tell you the water depth or distance underwater to structures such as wrecks or pinnacles (Vexilar LPS-1, HawkEye, Scubapro PDS). For larger vessels, there are a variety of depth sounders and "fish finders" that can aid the scientist in locating a site. Side-scan sonar can be used to search for large features such as pinnacles, canyons, or physical disturbances such as ice scour, and to map large areas (Heine, 1989; Anuskiewicz and Garrison, 1992). Researchers have also used side-scan sonar to locate gray whale feeding excavations in the soft-bottom sea floor (Oliver and Kvitek, 1984) as well as other physical and biological features (Kvitek et al., 1998).

Buoys

SURFACE BUOYS ARE PERHAPS THE BEST AND USUALLY the easiest method for relocating a site from the surface. However, they take some time to install correctly, and are subject to loss from storms, theft, entanglement in boat propellers, or mauling by marine animals (i.e. sea otters). Inexpensive surface floats can be made from plastic bottles, which can be filled with foam, or rubber tire tubes. High-density foam torpedo-shaped buoys resist wave motion and decrease the chances of entanglement in kelp. An additional buoy placed near the bottom will keep the line from dragging at low tide. Some people prefer to use subsurface buoys to cut down on theft, as well as to discourage curious boaters and divers from using it as a mooring. There is a trade-off between keeping it deep enough for other boaters to avoid catching it in their propeller, and being able to easily locate it from the surface (see Foster et al., 1985).

Buoys can be connected to the bottom with a variety of types of lines, cable, or chain. Temporary buoys can be as simple as a plastic bottle tied to a polypropylene line with a small weight on the bottom. For a more substantial and permanent buoy, strong line, chain, or steel cable should be used. Attachment to the bottom can be done with heavy weights such as engines, concrete bumpers, railroad track, and iron I-beam. To avoid chafing, a length of garden hose can be placed over the line and hardware. Alternatively, holes can be drilled into the substrate and eyebolts can be cemented into the rock. In sandy or soft-bottom areas, sand, earth, or fence anchors can be screwed into the bottom to provide a strong buoy attachment point. They can also be used to secure instruments or cages to soft bottom environments. They can be found at larger hardware stores, or at trailer or mobile home installation companies.

PHYSICAL MANIPULATIONS OF THE ENVIRONMENT

Underwater Marking

ONCE THE SURFACE LOCATION OF A STUDY SITE is established, it may be necessary to mark the site underwater as well. In rocky areas, pitons, boat nails, tent stakes, railroad spikes, cement (masonry) nails, or sharpened rebar can be driven into the substrate with a small 1.4 kg (3 lb) short-handled sledge hammer. It is often helpful to find a small crevice or crack to start the nail or other marker in. Various types of marking tags can be placed on the nail, such as electrical cable ties, colored plastic bicycle handle bar tape, vinyl roll flagging tape, cow or goat ear tags, or pieces of PVC. Many of these tags can be written on with permanent ink, etched or scribed with an electric Dremel tool, or burned with a soldering iron (see Foster et al., 1985). Multiple nails can be installed with a pneumatic air hammer, leaving a specified length exposed as a marker (Gittings et al., 1990).

Drilling

HOLES CAN BE DRILLED BY HAND IN ROCK using a hammer and star drill. The star drill bit is held with vice grip pliers, and is rotated slightly with each blow of the hammer. This can be time-consuming and tiring, and is not practical if a large number of holes need to be made. For numerous holes, especially in very hard rock, or for more permanent fastening, pneumatic tools can be used. A simple air-driven drill or impact hammer with a masonry bit can be fitted to operate off of a standard scuba regulator first stage, using the intermediate pressure of about 130 psi. Gittings et al. (1990) describe using a pneumatic impact wrench with a coring bit to drill 6.5 cm (2.6 in) holes, into which eyebolts were cemented using Portland Type II cement. A more heavy-duty drill, such as a Chicago Pneumatic Hand Drill (Model CP-9A or 9F) will drill faster and into harder substrate, but also uses more air. These drills also require considerable maintenance after every use, which includes partial disassembly,

FIGURE 4.1:
Underwater drilling into coral using an inexpensive pneumatic drill powered by a scuba cylinder. (photo courtesy of A. Gelber, PBS&J)

cleaning, and soaking in an oil, such as diesel fuel, to prevent corrosion. Underwater power tools can generate high noise levels, and care should be taken to protect the diver from detrimental sound fields (Hollien and Hollien, 1990).

Yund (pers. comm.) describes drilling holes in granite with self-drilling wall anchors. The brand shown here is sold under the name of Red Head, and the part is an S14 anchor. The wall anchor accepts a 1/4 inch bolt, or other fasteners can be epoxied into the anchor. Other manufacturers (including Rawl) sell virtually identical anchors. Industrial fastener supply houses should be well-stocked with these parts.

FIGURE 4.2:
Wall anchor, expander, and chuck.
(photo courtesy of P. Yund)

Air-driven drills can consume large quantities of air, especially at greater depths, and can be quite loud. After use in salt water, air hammers and drills should be soaked in fresh water, then have air and lubricant run through them.

Hydraulic systems can be more efficient, and are certainly quieter, but are relatively expensive, and are required to be linked to a control station on the surface (see Jaap et al. 1990 for details). Explosive tools (e.g. Ramset) can also be used to place nails or studs into rock.

Cement and Epoxy for Securing Hardware to the Substrate

THERE ARE A NUMBER OF DIFFERENT TYPES OF CEMENTS and epoxies available for use underwater. They are available at marine supply, swimming pool products stores, and many hardware stores. As with most glues and cements, they work best on clean substrata. They can be packed into holes that have been drilled, or onto clean substrate, especially in cracks and crevices. Jaap et al. (1990) have used four parts of Type II Portland cement with one part molding plaster. The cement is mixed with seawater and placed into plastic bags for use underwater. It is packed full into holes before an eyebolt or stake is placed in. Other researchers have used marine putties or underwater patching compound designed for swimming pools (see list at end of chapter; Coyer and Witman, 1990; Coyer et al., 1999).

One very popular epoxy is called Z Spar marine epoxy (A-788 Splash Zone Compound; Koppers, Inc.). This two-part epoxy is mixed on the surface, usually on a piece of plywood, and is useable for about one hour after mixing. It cures very hard after about 12 hours.

After epoxy or cement is placed on the substrate, some type of hardware can be inserted into the epoxy for mounting instruments or securing a cage to the bottom. Stainless steel hardware is preferred for resistance to corrosion. Hardware such as eyebolts, threaded rod, anchor wedges, or other fittings can be attached onto rock substrates (Foster et al., 1985; Goldberg and Foster 2002). There are also a number of different types of anchoring devices which can be placed into drilled holes or cracks and crevices, which have their own adhesives contained inside the unit. Other ingenious (but potentially expensive) products can be found at mountaineering or rock climbing stores.

TRAINING EXERCISE #2
Locating and Marking a Site on Surface and Underwater

Equipment needed: slates, compasses, GPS, buoys, line, sledge hammers, nails (appropriate for the substrate), epoxy, eyebolts, drill, air cylinders

Objective: To establish a study site, and mark its location on the surface and underwater, using a variety of methods

At the desired site, the entire group can participate in establishing line-ups, taking compass bearings, and using other navigational tools available to construct a surface map of the site. For the underwater exercises, the class can be divided into buddy pairs or groups of three. A thorough briefing must be given by the teaching staff prior to the dive. A "dry run" on land is often beneficial prior to the dive. If used, the epoxy should be mixed up on the surface prior to the dive. A staff member can assemble the drill at the study site (placing it at the boat anchor makes it convenient for all dive team members to find). Each student can practice drilling, hammering nails, and cementing eyebolts into the substrate with epoxy. A surface buoy and line can be rigged for future attachment to the eyebolt after the epoxy has set up. This is also a good opportunity for the class to relocate the site using the surface map they prepared earlier.

FIGURE 4.3: Practicing dry runs of exercises on land will often minimize problems and mistakes underwater. (photo courtesy of J. Pye)

TRAINING EXERCISE #3
Practice Using Tools Underwater (Multi-Station Exercise)

Equipment needed: Commonly used underwater tools (drill, epoxy, eye bolts, hammers, chisels, wire brushes, cages, lift bags)

Objective: Orientation and practice using underwater tools. This exercise may be substituted for or combined with Exercise #1

The class can be divided into buddy pairs. A thorough briefing must be given by the teaching staff prior to the dive, including the overall dive plan and the use and precautions of each piece of equipment. The group can help assemble all research equipment, and prepare the underwater epoxy mix. Underwater, various stations can be set up close to each other to allow the students to briefly visit each station and operate the equipment provided. It is helpful to have a staff member supervise each station.

The first station might have an underwater drill set up where the divers can practice drilling into the substrate. After each person drills or partially drills a hole, they can practice cementing in an eyebolt with underwater epoxy. Another station might have tools such as a hammer, chisel, and wire brush to clear the substrate of organisms. The divers could then practice affixing a cage to the substrate with epoxy. Another station could have a lift bag and a weight to lift. It is suggested that the weight be less than 23 kg (50 lb). A spare weight belt works well. Each buddy pair will attach the lift bag, fill it from an extra cylinder and regulator, and ascend at a normal rate to the surface. This should be done in shallow water (less than 10 m; 33 ft), and close supervision by a staff member is recommended. Additional stations can be added as required and might use such tools as hammers, nails, pitons, or metal detectors.

The group should also be instructed on proper care and maintenance of the equipment after the dive. A debriefing is helpful to discuss problems encountered and to solicit suggestions for improvement or modification of the method for particular circumstances.

Underwater Navigation and Mapping

EARLY FORMS OF UNDERWATER MAPPING USED an underwater theodolite, an underwater plane table, and lasers for surveying (Farrington-Wharton, 1976). Today, the simplest tools to use for underwater mapping include a slate, transect tape, depth gauge, compass, and inclinometer. Slates can be purchased at office supply stores, or constructed from aluminum or flat stock PVC. Data can be recorded directly onto the slate, which then must be transposed or photocopied later, or data sheets can be drawn on underwater paper or Mylar acetate sheets. Data sheets can be secured to the slate with a bar and wing nuts, with spring clips (which will rust), or with bungee cord or surgical tubing. Pencils can be tethered by string or tubing. Extra pencils should always be taken underwater (see Foster et al., 1985).

FIGURE 4.4:
Common mapping tools include slates, transect tapes, depth gauges, compasses, and cameras. (photo courtesy of J. Pye)

Transect tapes are available in a variety of types, lengths, and graduations. One popular model is an open reel fiberglass tape, in lengths up to 100 m (330 ft), with graduations in feet and inches on one side, and centimeters and meters on the other side (see Forestry Supply ref.). It is helpful to put a brass snap hook on the end of the tape to secure it to an anchor line, or other marker that signifies the beginning point of a transect. Transect tapes should be wound carefully to avoid twisting and tangling underwater, and should be thoroughly rinsed after the dive.

A standard underwater diving compass is usually sufficient for most mapping purposes. More accurate compasses are available from outdoor or surveying stores or catalogs. Compasses mounted on a swim board may be more accurate (see RJE International). Electronic compasses are now available for underwater use, and may provide more accurate and reliable navigation than analog versions (see True North Technologies). Some additional advantages of electronic compasses include not having to be held in a level position, the ability to store numerous courses, bearings, times, and declination angles, and different operational modes. An automatic return feature will calculate a return course to the starting point.

A diver navigation platform allows a diver to navigate underwater without beacons or surface support (DNC-250, RJE International). A diver's position is plotted using geodetic position, depth, heading, and velocity. The diver display includes distance and bearing to target and depth.

A diver-worn wrist instrument shows latitude and longitude, a detailed map of underwater terrain, current heading and depth, position of nearby divers (within about 100 yds/90 m), and direction back to vessel (Navimate). The system requires a small transmitter to be hung from a boat or buoy.

Inclinometers (Suunto; see Forestry Suppliers cat.) can be used underwater to measure heights of reefs and pinnacles, to measure the slope of rocks or sediment ripples, and to measure vertical angles of the substrate. They can read in percent and/or degree scales.

An example of a detailed mapping exercise is the Clifton S. Perry Memorial Artificial Reef Monitoring Program in Florida. The project consisted of mapping old bridge pilings that were dropped for the artificial reef. Figure 4.5 shows the final mapped diagram of an area of the reef (Krug, pers. comm.). The divers started at benchmark #4, and worked off of a 30°/210° azimuth. In two dives, the structures in the area were mapped, and the distances from Benchmark #4 were noted. This sketch was then used as raw data and placed into a computer program that mapped all of the pilings around the reef.

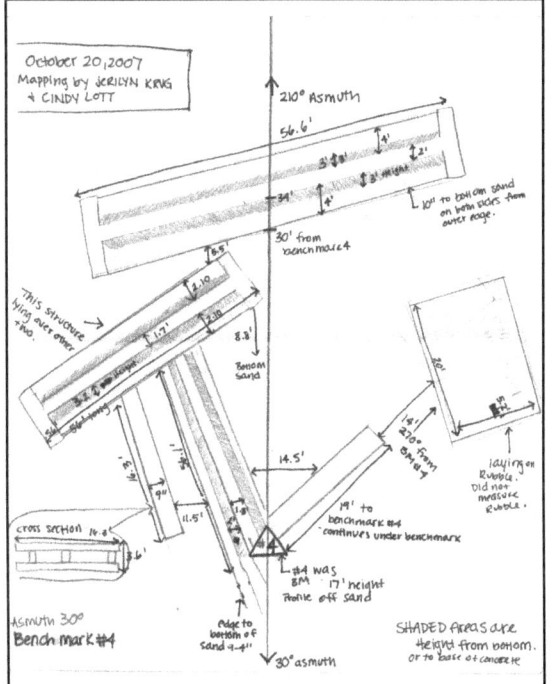

Figure 4.5:
Sample map made from underwater measurements by divers.
(photo courtesy of J. Krug)

TRAINING EXERCISE #4
General Underwater Mapping

Equipment needed: slate, transect tape, depth gauge, compass, and inclinometer.

Objective: To construct a map of underwater terrain, including compass bearings, and three dimensions of reefs and other patches of substrate. Major features, such as number of kelp plants, large coral outcrops, or large sponges can also be counted and/or measured.

Depending upon the number of students available, the staff will select a reasonable sized area underwater to map, such as 10 m (33 ft) x 10 m (33 ft) or 20 m (66 ft) x 20 m (66 ft). Transect tapes, or fixed-lengths of line can be laid out underwater to delineate the area. The diving buddy pairs can be assigned the specific tasks as outlined in the objectives above. The Lead Diver will be responsible to check the completeness and adequacy of the data before leaving the site. Dive groups can rotate tasks during the dive, or on subsequent dives. The data can be pooled in class for evaluation and write-up. A debriefing is helpful to discuss problems encountered and to solicit suggestions for improvement or modification of the method for particular circumstances or environments.

The US EPA dive team in the Pacific Northwest maps and records the exact locations of underwater resources and hazards in low visibility conditions, using relatively inexpensive technology (Siwiec et al., 2008). Divers record objects of interest using digital photography and recreational GPS units. The GPS device is towed by the divers in a surface float directly above them. Commercial software is used to relate the digital photos and time-stamped positions to be used on a map or Geographic Information System (GIS). For a complete description of the survey technique, see Siwiec et al., 2008.

A rapid and accurate mapping technique using scuba divers and GPS has been used to map seagrass communities in the Mediterranean Sea (Sgorbini et al., 2002). They used a GPS "intelligent" buoy that contained an underwater pinger, four buoys, and a receiving deck unit. The system has a reported accuracy of 0.5 m (1.5 ft.).

There are a number of electronic sonar and navigation systems that can be used by divers. One simple system called the DiveTracker Sport uses an underwater transmitter that can be placed on a study site or other underwater location for an extended period of time. A small diver-held receiver, with a range of up to 1300 m (4000 ft), is used to guide the diver to the marked location (see Desert Star Systems ref.). Some systems use multiple frequencies, which allow for the marking and relocation of multiple objects or sites within the same area (see also RJE International ref.).

For advanced underwater mapping and precision surveying, a system called the Aqua-Map™ is available (Desert Star Systems). Rather than using time-consuming and bulky survey lines, this system uses baseline stations that transmit SONAR pulses that are picked up by diver-held receivers as they swim in an area. The system even guides divers to complete a map of an area as large as 500m x 500m, with sub-meter accuracy. Electronic observations can be recorded by pressing a button, associating it with position, depth, and time, with an accuracy of tenths of centimeters.

Larger areas can be searched and roughly mapped by divers using DPVs or being towed by a boat. Some method of recording data on an audio tape would be most useful here (see Communications, Ch. 3). Additionally, side-scan sonar can be used to locate specific habitats or environments. A typical system uses a 100 kHz frequency (e.g. EGG Model 259-4; Klein) and a tow fish, which contains the underwater transducers. This fish is towed at a constant speed behind a vessel at a depth just above the bottom, and recordings of a 50 m (165 ft) swath on either side of the tow fish can be made (Oliver and Kvitek, 1984; Heine, 1989; Anuskiewicz and Garrison, 1992).

FIGURE 4.6
Diver using AquaMap™ sonar navigation system. (photo courtesy of T. Gray)

Scientific divers may also want to mark a site electronically for relocation purposes. There are pingers designed to be placed on the bottom, or in mid-water, that give off a signal to a diver-held receiver (RJE International, Desert Star systems). Acoustic release devices (e.g. Digicourse) are often used in deeper water to retrieve instruments.

The National Park Services Submerged Resources Center has been tracking movements on the USS Arizona using a kinematic GPS unit (Trimble) with a tripod that is set up on points on the wreck. These points are evaluated to determine any movement and are accurate to under a centimeter in three dimensions.

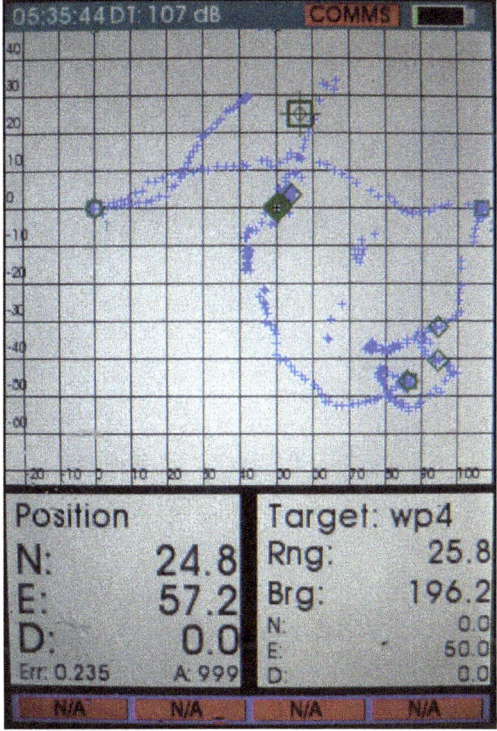

FIGURE 4.7
Sample readout of AquaMap™ diver terminal. (photo courtesy T. Gray)

FIGURE 4.8:
Scientific diver adjusts the GPS unit that tracks movements on the **USS Arizona**.
(photo courtesy of B. Seymour)

Figure 4.9:
A scientific diver searches an area with a metal detector.
(photo courtesy of D. Kesling)

Scientific divers have used underwater metal detectors for such tasks as locating ancient artifacts or animals tagged with metal tags. Boat-towed units are available that have high sensitivity and a wide range of detection (see J. W. Fishers ref.).

Equipment Manufacturers and Suppliers

ACSA Underwater GPS
9 Europac
13590 Meyreuil
France
Phone: 33 (0) 442 58 54 52
www.underwater-gps.com

Chicago Pneumatic (underwater drills)
3700 E. 68th Ave.
Commerce City, CO 80022
(800) 760-4049
www.cp.com

Deep Ocean Engineering (ROV's)
1431 Doolittle Dr.
San Leandro, CA 94577
(510) 562-9300
www.deepocean.com

Desert Star Systems (underwater positioning, navigation systems)
3261 Imjin Road
Marina, CA 93933
(831) 384-8000
www.desertstar.com

ION (marine imaging systems)
5200 Toler St.
Harahan, LA 70123
(504) 733-6062
www.iongeo.com

EdgeTech (hydrographic instruments; side scan sonar)
4 Little Brook Rd.
West Wareham, MA 02576
(508) 291-0057
www.edgetech.com

Forestry Suppliers, Inc. (compasses, transect tapes, inclinometers)
205 W. Rankin St.
Jackson, MS 39284
(800) 647-5368
www.forestry-suppliers.com

J.W. Fishers Mfg., Inc. (sonar, underwater metal detectors, cameras, video)
1953 County St.
E. Taunton, MA 02718
(800) 822-4744
www.jwfishers.com

Klein Associates Inc. (sonar)
11 Klein Dr.
Salem, NH 03079
(603) 893-6131
http://www.kleinsonar.com

Koppers, Inc. (Z Spar Splash Zone Epoxy)
P.O. Box 911041
Commerce, CA 90011

Navimate (GPS for divers)
19215 Parthenia St., Suite A
Northridge, CA 91324
(818) 773-2000
www.navimate.com

Recreation Supply Co. (underwater epoxy, patching)
515 Airport Rd.
Bismarck, ND 58504
(701) 222-4860
www.recsupply.com

RJE International (sonar, diver navigation, pingers, and receivers)
15375 Barranca Parkway, Suite B-107
Irvine, CA 92618
(949) 727-9399
www.rjeint.com

True North Technologies Corp. (electronic compasses)
Two Clock Tower Place
Suite 335
Maynard, MA 01754
(978) 897-5400
www.tntc.com

West Marine (epoxy, waterproof fittings)
www.westmarine.com

References

Anuskiewicz, R.J. and E.G. Garrison. 1992. Underwater archaeology by Braille: Survey methodology and site characterization modeling in a blackwater environment - a study of a scuttled confederate ironclad, C.S.S. Georgia. In: Cahoon, L.B. (ed.), Diving for Science...1992, Proceedings of Amer. Acad. Underwater Sci. pp. 1-12.

Carleton, J.H. and P.W. Sammarco. 1987. Effects of substratum irregularity on success of coral settlement: Quantification by comparative geomorphological techniques. Bull. Mar. Sci. 40: 85-98.

Coudeville, J.M. and H. Thomas. 1998. A primer: Using GPS underwater. Sea Technology 39(4): 31-34.

Coyer, J., and J. Witman. 1990. The Underwater Catalog: A Guide to Methods in Underwater Research. Shoals Marine Laboratory, Cornell University, NY. 72 pp.

Coyer, J., D. Steller and J. Witman. 1999. The Underwater Catalog: a guide to methods in underwater research. 2nd Edition, Shoals Marine Laboratory, Cornell Univ., Ithaca, NY. 151 pp.

Fager, E., A Fleichsig, R. Ford, R. Clutter, and R. Ghelardi. 1966. Equipment for use in ecological studies using scuba. Limnol. and Oceanogr. 11(4): 503-509.

Foster, M.S., T.A. Dean, and L.E. Deysher. 1985. Subtidal Techniques, pp. 189-232. In: Littler, M.M. and D.S. Littler (eds.). Handbook of Phycological Methods. Ecological Field Methods: Macroalgae. Cambridge University Press. 617 pp.

Gittings, S.R., K.J.P. Deslarzes, B.S. Holland, and G.S. Boland. 1990. Ecological monitoring on the Flower Garden Banks: Study design and field methods. In: W.C. Jaap, (ed.), Diving for Science...1990, Proceedings of Amer. Acad. Underwater Sci. pp. 107-118.

Goldberg, N.A. and M.S. Foster. 2002. Settlement and post-settlement processes limit the abundance of the geniculate coralline alga *Calliarthron* on subtidal walls. J. Exp. Mar. Biol. Ecol. 278: 31-45.

Heine, J.N. 1989. Effects of ice scour on the structure of sublittoral marine algal assemblages of St. Lawrence and St. Matthew Islands, Alaska. Mar. Ecol. Progr. Ser. 52: 253-260.

Hollien, H. and P.A. Hollien. 1990. Effects of diver tool use on diver hearing. In: W.C. Jaap, (ed.), Diving for Science...1990, Proceedings of Amer. Acad. Underwater Sci. pp. 163-178.

Jaap, W.C., J.L. Wheaton, and K.B. Donnelly. 1990. Materials and methods to establish multipurpose, sustained, ecological research stations on coral reefs at Dry Tortugas. In: W.C. Jaap, (ed.), Diving for Science...1990, Proceedings of Amer. Acad. Underwater Sci. pp. 193-203.

Kvitek, R.G., K.E. Conlan, and P.J. Iampietro. 1998. Black pools of death: Hypoxic, brine-filled ice gouge depressions become lethal traps for benthic organisms in a shallow Arctic embayment. Mar. Ecol. Progr. Ser. 162: 1-10.

Medard, J. 1997. Underwater positioning for archaeology and topographical needs. Sea Technology, vol. 38, no. 7, pp. 31-33.

Oliver, J.S. and R.G. Kvitek. 1984. Side-scan sonar records and diver observations of the gray whale (*Eschrichtius robustus*) feeding grounds. Biol. Bull. 167: 264-269.

Sgorbini, S., A. Peirano, S. Cocito, and M. Morgigni. 2002. An underwater tracking system for mapping marine communities: an application to *Posidonia oceanica*. Oceanologica Acta 25: 135-138.

Siwiec, T., S. Sheldrake, A. Hess, D. Thompson, L. Macchio, and P.B. Duncan. 2008. Survey technique for underwater digital photography with integrated GPS location data. In: Brueggeman, P. and N.W. Pollock, (eds.), Diving for Science 2008, Proceedings of Amer. Acad. Underwater Sci. pp. 159-166.

Witman, J.D. 1987. Subtidal coexistence: Storms, grazing, mutualism, and the zonation of kelps and mussels. Ecol. Monogr. 57: 167-187.

CHAPTER 5

ARCHAEOLOGY

ARCHAEOLOGY IS THE STUDY OF HISTORICAL SOCIETIES and relics, which include artifacts, tools, vessels, and other remains. Archaeologists locate, draw, excavate, and recover objects with historical significance. A considerable amount of underwater archaeology is conducted on historic shipwrecks found in riverine, estuarine, and coastal environments. These areas usually have very poor water visibility conditions, as well as strong currents and ship traffic, making archaeological research such as mapping and excavation extremely difficult (Anuskiewicz and Garrison, 1992). Useful tools for this type of research include a magnetometer survey, tight-beam bathymetric survey, and side-scan sonar surveys. For a general review of underwater archaeological techniques see Dumas (1962) and Bass (1970).

FIGURE 5.1:
Divers on an archaeological project often use technical diving equipment, such as double cylinders and decompression stage cylinders, especially in deeper water.
(photo courtesy of D. Kesling)

Magnetometers record magnetic anomalies, which in conjunction with navigational fixes are used to produce a contour map. Computer software is available to process the large amount of data that can be collected. Detailed bathymetric maps can be overlaid to produce a better representation on the shipwreck. Three-dimensional graphics programs allow the researcher to "view" the shipwreck from a variety of angles (Anuskiewicz and Garrison, 1992). Side-scan sonar is also useful for an overall view of the entire shipwreck. Pulse length and frequency can be adjusted to produce images of varying resolutions.

An important part of any archaeological study is the survey. After that, an excavation may take place, where features or relics may be uncovered or collected. Many underwater sites, such as those in estuaries or rivers, have poor visibility due to fine sediments or water quality, which limit the amount of light penetration. They may also be located in areas of strong water flow such as currents, tides, or surf, and may also have significant entanglement possibilities. Pool training may be useful to simulate working under these conditions (see Ch. 3, Zero-Visibility Diving).

Site Survey and Underwater Mapping

MANY ARCHAEOLOGICAL SITES REQUIRE A TOPOGRAPHICAL survey and map showing the location of artifacts and important reference points. This can be accomplished by using simple tools such as tape measures, compasses, depth gauges, and slates (see Figure 4.6 for a sample map).

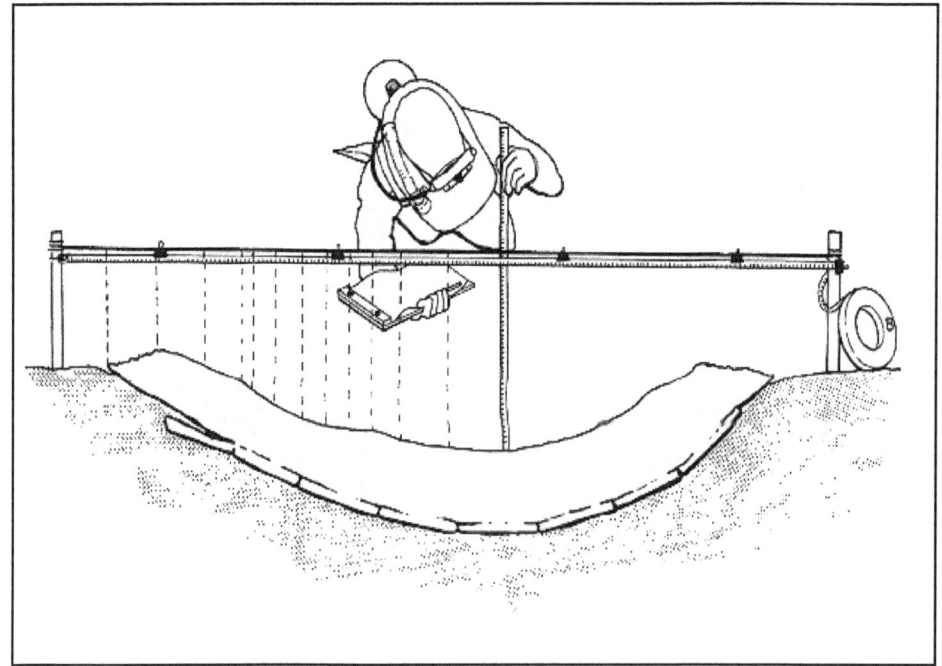

FIGURE 5.2:
Diver recording vertical offset measurements from a horizontal baseline.
(drawing courtesy of S. Sellers)

A map is a way to represent objects in a two or three-dimensional space. A site map is a two-dimensional plan view that looks down from above. The third dimension provides an elevational aspect, where cross sections show the vertical components of a site.

The first step in mapping is to establish fixed points, called datums, which are permanent locations that measurements are taken from. A baseline is run between datums, which facilitates mapping and creates a grid zone of the site. Horizontal offsets are measurements taken at right angles to the baseline. Vertical offsets are taken from a level baseline using a plumb bob and scaling rod. In very low-visibility conditions, divers can use "Brody Bags," which are clear Ziploc bags filled with water. The bags can be placed over a measuring tape, and when a dive light is shined on it, the numbers can be more easily read. Another ingenious measuring device consists of a tape measure that is sandwiched between two layers of clear Lexan sheets, with a clear semicircular Lexan viewing tube filled with clear water cemented to the Lexan sheet. An underwater light mounted next to the tube illuminated the tape sufficiently for measurements to be read from the tape and relayed to the surface for recording (Cantelas and Rodgers, 1997).

Another step in mapping is called triangulation, where distance measurements from two or more datums are used to map the precise location of a target item. This method is accurate over great distances, but must be in the same horizontal plane.

A diver-operated computer has been developed for underwater archaeological three-

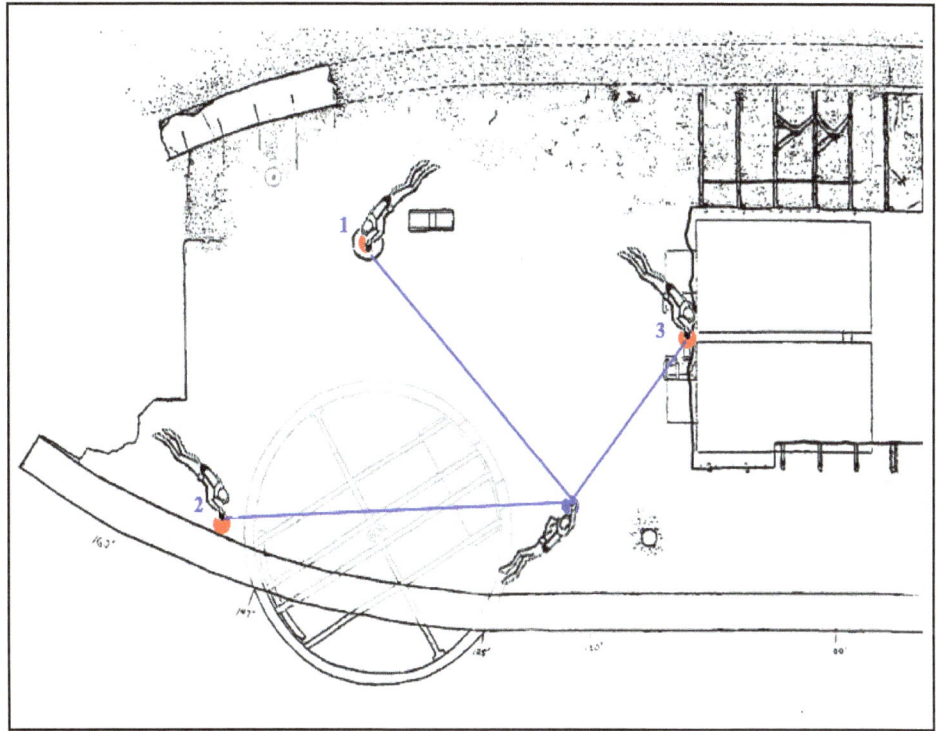

FIGURE 5.3:
Divers using the triangulation method to map the location of a target.
(drawing courtesy of S. Sellers)

dimensional mapping purposes (Carter and Covill, 1986). It has a programmable calculator and display in a waterproof housing, with a fiber optic wand that scans preset bar codes for data entry. The diver can enter up to 200 artifact types and x, y, z coordinates into the calculator, and download them later to a personal computer.

There are also electronic mapping systems consisting of reference points marked with data receivers, and a diver-carried "mouse" is used to click on areas to be mapped as the divers cover a site (White, 1987).

More traditional surveying and mapping methods include the use of a Peterson's wheel to measure elevations, transect tapes for triangulation, plane tables, and alidades. These methods have limitations based on water visibility, size of the area to be mapped, and the available communications equipment for the divers. Another method, used by Merifield and Rosencrantz (1966) and Rosencrantz (1975), is useful for determining three-dimensional positions. It uses measurements from perimeter reference stakes, and angles can be calculated later through triangulation.

If the water conditions are appropriate, photography or videography methods may be used to map the site and provide a record of important items. Traditionally, digital still photos were taken and later stitched together to form a photomosaic image of the site. The Wisconsin Historical Society, in conjunction with the Advanced Imaging and Visualization Laboratory at Woods Hole Oceanographic Institution, used video cameras that were adapted to low light levels and 250 watt HID lamps to image deep shipwrecks (Thomsen and Meverden, 2003). The camera and lights were mounted on a diver propulsion vehicle

FIGURE 5.4:
Divers measuring deck winch on a shipwreck.
(photo courtesy of D. Kesling)

(DPV) pointing downwards. Divers would swim 6-9 m (20-30 ft) above the wreck, keeping the camera pointed at a perpendicular angle to the bottom, while "mowing the lawn" to capture the entire structure. A fiber optic cable relayed images to a topside computer. Three-dimensional models of wrecks with a precision of 1 cm are being developed. Images for a complete photomosaic can be made in a matter of a few minutes dive time. See Thomsen and Meverden (2003) for a more complete description of the technique.

FIGURE 5.5:
Video documentation of an archaeological site.
(photo courtesy of D. Kesling)

FIGURE 5.6:
Video camera and HID lights mounted on a DPV for the creation of a photomosaic.
(photo courtesy of T. Thomsen, Wisconsin Historical Society)

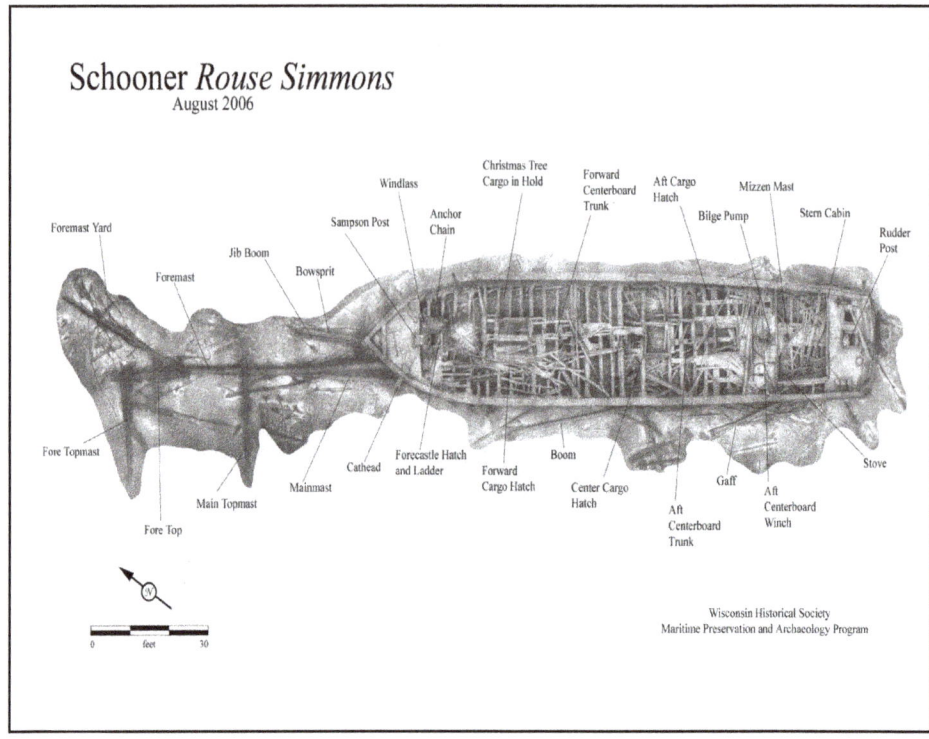

FIGURE 5.7:
Photomosaic of Schooner *Rouse Simmons* showing important features.
(courtesy of T. Thomsen, Wisconsin Historical Society)

TRAINING EXERCISE #5
Archaeology Underwater Mapping in Pool

Equipment needed: slates, transect tapes, depth gauges, compasses, underwater cameras.

Objective: To construct a map of an underwater "site," including compass bearings, and three dimensional measurements. Major features, such as a vessel "bow" and "stern" (trash can or white buckets), cleats, anchor, etc. can also be counted and/or measured.

The staff can set up an underwater section of the pool to map, such as 10 m (33 ft) x 10 m (33 ft) or 20 m (66 ft) x 10 m (33 ft). Weighted lengths of line can be laid out underwater to delineate the area. The diving buddy pairs can be assigned the specific tasks as outlined in the objectives above. The lead diver will be responsible to check the completeness and adequacy of the data before leaving the site. Dive groups can rotate tasks during the dive, or on subsequent dives. If done at night, the pool lights can be dimmed or shut off in order to simulate low-visibility conditions. The data can be pooled in class for evaluation and write-up. A debriefing is helpful to discuss problems encountered and to solicit suggestions for improvement or modification of the method for particular circumstances or environments.

FIGURE 5.8:
Diver taking notes on historical shipwreck.
(photo courtesy of B. Seymour)

FIGURE 5.9:
A surface-supplied diver operates a water induction dredge.
(photo courtesy of S. von Arbin)

Excavation

MANY ARCHAEOLOGICAL sites are covered with sediment, which must be removed to allow investigations to take place. One simple method that can be used if sediment is light is hand-fanning. If the sediment layer is thick, there are methods that can move a significant amount of sediment rapidly. One such method is the water induction dredge. A water pump supplies high-pressure water into a hose (diameter 4 in, 10 cm), which causes an induced suction at the working end of the hose. This type of dredge has been used in springs and sinks to locate lithic artifacts (Anuskiewicz and Dunbar, 1993). Marine sediments deposited in the spring, especially in crevices, are removed with the stream of water.

Another method to remove sediment is the use of an airlift. Pressurized air is introduced into the bottom of a tube (usually pvc pipe) from either a surface compressor or scuba cylinder. As the air rises up the pipe it expands, which produces a suction effect at the bottom of the pipe (see Ch. 7 for details).

The National Park Service has an active group of diving archaeologists, the Submerged Resources Center (formerly the Submerged Cultural Resources Unit), which uses underwater archaeology as a research and management tool (Lenihan, 1985; Bozanic, 2007; http://www.nps.gov/applications/submerged/projsub.cfm). This group conducts projects such as historic shipwreck surveys and mapping in a variety of environments such as marine, rivers,

CHAPTER 5 ARCHAEOLOGY **89**

reservoirs, sinkholes, and caves. Their mission is to enhance public appreciation and access to submerged resources.

Tagging and Retrieval of Artifacts

HISTORICAL ARTIFACTS CAN BE SUBJECT TO DEGRADATION due to corrosion and natural decay. Wooden structures and metal objects in warm seawater are particularly subjected to decay, including biological encrustation. Submerged cultural resources generally are better preserved in colder waters. Neyland (2005) reported that a Confederate submarine that sunk in 1864 was very well-preserved due to being filled with fine silt, which created an anaerobic environment that slowed the degradation of iron, wood, and leather.

Examples of Scientific Archaeological Diving

STANTON ET AL. (2007) REPORT using closed-circuit rebreathers in a forensic study of the *Rouse Simmons* shipwreck in 165 ffw (50 mfw) in Lake Michigan. Measurements taken on the 160 ft (49 m) three mast wooden schooner included the length, circumference, and diameter of the spars, fittings, and standard rigging, resulting in the drawing of a general site plan. A very detailed description of the shipwreck was constructed after two five-day dive trips.

The Maritime Studies Program at East Carolina University has conducted numerous studies of shipwrecks in the Great Lakes (Corbin and Rodgers, 2003), British Virgin Islands (Gleason, 2003), and North Carolina coast. Masters and doctoral degrees in Maritime History and Nautical Archaeology, as well as in Coastal Resources Management are offered (Runyan, 2003).

Several major World War I warships sunk in Scotland have been catalogued and mapped (Forbes, 2003). Initial base maps

FIGURE 5.10:
A diver collects an historical artifact.
(photo courtesy of D. Kesling)

FIGURE 5.11:
A diver tagging archaeological contents.
(photo courtesy of S. von Arbin)

FIGURE 5.12:
Diver raises a fragile hull timber using a lifting box.
(photo courtesy of S. Sellers)

were constructed using remote sensing techniques, and a subsequent photomosaic was constructed using scuba diver and surface supplied diver images. The data will be incorporated into an updated divers guide and for interpretative displays at local museums.

A Confederate submarine, the *H.L. Hunley*, was discovered off the coast of South Carolina, and subsequently recovered to protect it from looting and to identify and bury the military crew members (Neyland, 2005). The vessel was found to be quite different in construction from what was previously thought. Many hundreds of historical artifacts have been collected.

The RPM Nautical Foundation has been mapping the Albanian seafloor to depths of 100 m (330 ft) to locate and document submerged cultural resources (Smith, 2009). Divers surveyed targeted areas to identify discovered artifacts and establish site boundaries. Partial photomosaics were also constructed.

FIGURE 5.13:
Cave diver measures pre-historic sloth bones.
(photo courtesy of U. Kunz)

NOAA's National Marine Sanctuaries Maritime Heritage Program conducts diving projects associated with historical, cultural, and archaeological resources in the sanctuaries. Two particular projects, in the Papahanaumokuakea Marine National Monument in the Hawaiian Islands and the Thunder Bay National Marine Sanctuary, highlight some of the recent work. In Hawaii, two 19th century British whaling ships have been documented. In Lake Huron, most of the shipwrecks that lie in deeper water are very well preserved. The sanctuary is working with university partners to more fully understand how the chemical, biological and physical conditions are affecting the historical shipwrecks (Casserley, 2009).

The *Monitor* National Marine Sanctuary, along with other institutions, recently conducted baseline data-gathering dives on three German U-boats off North Carolina (Hoyt, 2009). The sites were documented with video and digital photographs, a detailed site plan, photomosaics, and visual observations. The overall goal was to augment the historical significance of the site and vessels, and enhance their nomination to the National Register of Historical Places.

Equipment Manufacturers and Suppliers

ACSA Underwater GPS
9 Europac
13590 Meyreuil
France
Phone: 33 (0) 442 58 54 52
www.underwater-gps.com

Chicago Pneumatic (underwater drills)
3700 E. 68th Ave.
Commerce City, CO 80022
(800) 760-4049
www.cp.com

Deep Ocean Engineering (ROV's)
1431 Doolittle Dr.
San Leandro, CA 94577
(510) 562-9300
www.deepocean.com

Desert Star Systems (underwater positioning, navigation systems)
3261 Imjin Road
Marina, CA 93933
(831) 384-8000
www.desertstar.com

ION (marine imaging systems)
5200 Toler St.
Harahan, LA 70123
(504) 733-6062
www.iongeo.com

Forestry Suppliers, Inc. (compasses, transect tapes, inclinometers)
205 W. Rankin St.
Jackson, MS 39284
(800) 647-5368
www.forestry-suppliers.com

J.W. Fishers Mfg., Inc. (sonar, underwater metal detectors, cameras, video)
1953 County St.
E. Taunton, MA 02718
(800) 822-4744
www.jwfishers.com

Klein Associates Inc. (sonar)
11 Klein Dr.
Salem, NH 03079
(603) 893-6131
http://www.kleinsonar.com

Navimate (GPS for divers)
19215 Parthenia St., Suite A
Northridge, CA 91324
(818) 773-2000
www.navimate.com

RJE International (sonar, diver navigation, pingers, and receivers)
15375 Barranca Parkway, Suite B-107
Irvine, CA 92618
(949) 727-9399
www.rjeint.com/

True North Technologies Corp. (electronic compasses)
Two Clock Tower Place
Suite 335
Maynard, MA 01754
(978) 897-5400
www.tntc.com

West Marine (epoxy, waterproof fittings)
www.westmarine.com

References

Anuskiewicz, R.J. and E.G. Garrison. 1992. Underwater archaeology by Braille: Survey methodology and site characterization modeling in a blackwater environment - a study of a scuttled confederate ironclad, *C.S.S. Georgia*. In: Cahoon, L.B. (ed.), Diving for Science...1992, Proceedings of Amer. Acad. Underwater Sci. pp. 1- 12.

Bass, G. 1970. Archaeology underwater. Pelican Books, Baltimore, MD.

Cantelas, F.J. and B.A. Rodgers. 1997. Tools, techniques, and zero visibility archaeology. In: Maney, E.J., Jr., and C.H. Ellis, Jr. (eds.), Diving for Science 1997, Proceedings of Amer. Acad. Underwater Sci. pp. 29-40.

Carter, J. and J. Covill. 1986. Evaluation of a diver-operated computer as a tool for underwater archaeology. In: Mitchell, C.T. (ed.), Diving for Science...86, Proceedings of Amer. Acad. Underwater Sci. pp. 239-248.

Casserley, T.R. 2009. Torrid seas to icebound lakes: Shipwreck investigations within NOAA's National Marine Sanctuaries. In: Pollock, N.W. (ed.), Diving for Science 2009, Proceedings of Amer. Acad. Underwater Sci., Sea Grant Publ. No. CTSG-10-09. pp. 86-90.

Corbin, A. and B.A. Rodgers. 2003. Wrecked, abandoned, and re-used: Archaeological exploration on the Great Lakes. In: Norton, S.F. (ed.), Diving for Science 2003, Proceedings of Amer. Acad. Underwater Sci. pp. 26.

Coudeville, J.M. and H. Thomas. 1998. A primer: Using GPS underwater. Sea Technology 39(4): 31-34.

Dumas, F. 1962. Deepwater Archaeology. Routledge and Kegan Paul, London.

Forbes, B. 2003. ScapaMap: The Scapa flow maritime archaeology project. In: Norton, S.F. (ed.), Diving for Science 2003, Proceedings of Amer. Acad. Underwater Sci. pp. 28-36.

Gleason, K. 2003. Compelled to run his majesty's ship ashore: The story of the HMS *Santa Monica* as historical, cultural and environmental resource. In: Norton, S.F. (ed.), Diving for Science 2003, Proceedings of Amer. Acad. Underwater Sci. pp. 38.

Hollien, H. and P.A. Hollien. 1990. Effects of diver tool use on diver hearing. In: W.C. Jaap, (ed.), Diving for Science 1990, Proceedings of Amer. Acad. Underwater Sci. pp. 163-178.

Hoyt, J.C. 2009. 2008 battle of the Atlantic survey methodology. In: Pollock, N.W. (ed.), Diving for Science 2009, Proceedings of Amer. Acad. Underwater Sci., Sea Grant Publ. No. CTSG-10-09. pp. 66-74.

Lenihan, D.J. 1985. Underwater archeological research in the National Park Service. In: Mitchell, C.T. (ed.), Diving for Science...85, Proceedings of Joint International Scientific Diving Symposium, Amer. Acad. Underwater Sci. pp. 165-168.

Medard, J. 1997. Underwater positioning for archaeology and topographical needs. Sea Technology, Vol. 38, No. 7, pp. 31-33.

Merifield, P. and D. Rosencrantz. 1966. A simple method for surveying a small area underwater. Limnol. and Oceanogr. 11(3): 408-409.

Neyland, R.S. 2005. Underwater archaeology and the Confederate submarine *H.L. Hunley*. In: Godfrey, J.M. and S.E. Shumway (eds.), Diving for Science 2005, Proceedings of Amer. Acad. Underwater Sci. pp. 61-73.

Rosencrantz, D. 1975. Photography in archaeological research. In: Hays, Jr., E. (ed.), Underwater photography and photogrammetry. University of New Mexico Press, Albuquerque, NM.

Runyan, T. 2003. Shipwrecks and students: discovering maritime heritage. In: Norton, S.F. (ed.), Diving for Science 2003, Proceedings of Amer. Acad. Underwater Sci. pp. 75.

Siwiec, T., S. Sheldrake, A. Hess, D. Thompson, L. Macchio, and P.B. Duncan. 2008. Survey technique for underwater digital photography with integrated GPS location data. In: Brueggeman, P. and N.W. Pollock, (eds.), Diving for Science 2008, Proceedings of Amer. Acad. Underwater Sci. pp. 159-166.

Smith, D.M. 2009. Submerged cultural resource discovery in Albania: Surveys of ancient shipwreck sites. In: Pollock, N.W. (ed.), Diving for Science 2009, Proceedings of Amer. Acad. Underwater Sci., Sea Grant Publ. No. CTSG-10-09. pp. 19-22.

Stanton, G., K. Meverden, T. Thomsen, and J. Garey. 2007. Closed-circuit rebreathers in the forensic study of the *Rouse Simmons* shipwreck. In: N.W. Pollock and J.M. Godfrey (eds.), Diving for Science 2007, Proceedings of Amer. Acad. Underwater Sci. pp. 89-99.

Thomsen, T. and K. Meverden. 2003. Video capture, shipwreck photomosaics. Advanced Diver, 24: 56-59.

White, J.S. 1987. Mapping the *Isabella* shipwreck. In: M.A. Lang (ed.), Coldwater Diving for Science...1987, Proceedings of the American Academy of Underwater Sciences. pp. 279-284.

chapter 6

MEASURING PHYSICAL FACTORS

Research Diving Tools

SCIENTIFIC DIVERS GENERALLY USE SIMPLE TOOLS to conduct research underwater. Given the expense and remote location of many diving research projects, tools need to be rugged and function reliably. Many innovative techniques and materials have been invented by industrious researchers. Various types of coring devices, sediment collectors, mapping devices, inclinometers, and devices for measuring light, color, water visibility, sound, temperature, water motion, salinity, and nutrients will be discussed below.

Before selecting a tool or technique to use in underwater research, one must critically evaluate its utility for the objectives being proposed. This includes a search of the literature on the questions being asked. A number of relevant papers are listed in the reference section at the end of this chapter. In some cases, additional training in the use of these tools will also be required. A general review of tools and techniques for near shore oceanographic instrument installations can be found in Spencer (1989). Check with your diving safety officer or diving supervisor for suggestions or training before using most tools underwater.

FIGURE 6.1:
Pool practice with tools, fittings, and equipment to be used in the field can save valuable time on-site.
(photo courtesy of J. Auer)

Geological Measurements and Collections

COLLECTION OF GEOLOGICAL SAMPLES UNDERWATER can range from using a simple hammer, chisel, or pry bar and collection bag, to an elaborate airlift dredge that sends material to the surface. Early accounts of using scuba diving for geological collections and measurements can be found in Dill and Shumway (1954) and Fisher and Mills (1952). Divers can be used to visually evaluate areas of specific sediment types, and sample the sediments *in situ*, ensuring that representative and relatively undisturbed samples are taken. An ingenious system used to generate geoacoustical waves and simulate seismic waves has been developed, using a spear gun that shoots into a cast iron plate (Akal, 1985). A penetrometer, which is a calibrated spring-loaded pole, can be used to measure the consolidation of the sediments.

For sediment collection, coring devices of various sizes have been utilized for grain size analysis or stratigraphy determinations. An inexpensive method uses metal cans with plastic lids. The bottom of the can should be removed and replaced with a fine mesh screen, which allows for water to escape as the can is pushed into the sediment (open side down, mesh side up). The diver must then push the lid under the lip of the can and seal it, before removing the core from the substrate. Diver-held piston cores 6.5 cm to 15 cm (3 to 6 in.) are constructed from PVC, and are designed to collect a complete and undisturbed sediment sample (see Dornblaser et al., 1989 for detailed explanation). Wave-transported sediment has been collected in containers with a slit in the top, buried in the troughs of sand ripples (Cood and Gorsline, 1972). Small Ekman grabs and box cores can be manually inserted into the substrate and tripped by divers, ensuring the grab takes a proper and complete sample (Harper et al., 1989). A collection of sediment cores, or rock samples can be very heavy. Lift bags can be utilized to facilitate transferring the samples to the surface.

A diver-operated stainless steel box corer (15 cm [6 in.]

FIGURE 6.2:
A penetrometer is used to measure the consolidation of sediments.
(photo courtesy of J. Heine)

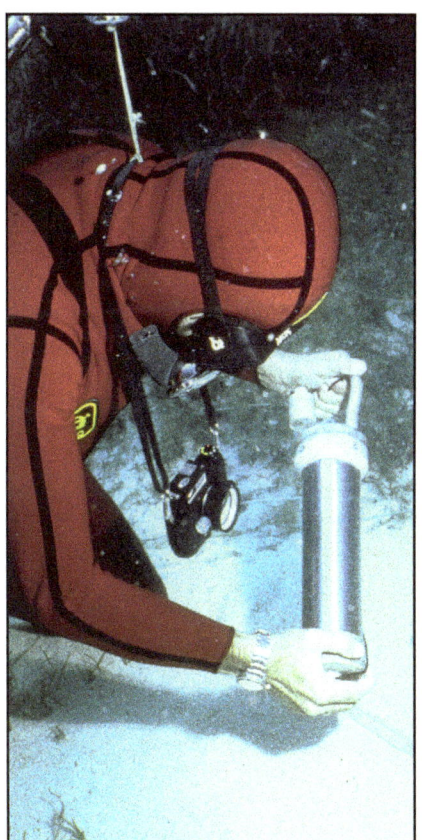

FIGURE 6.4:
A core rack with core cylinders and rubber stoppers. (photo courtesy of S. Sheldrake)

FIGURE 6.3:
A diver places a cap on a core sampler.
(photo courtesy of F. Cinelli)

FIGURE 6.5:
Divers operating a large hydraulic core sampler.
(photo courtesy of D. Kesling)

wide by 30 cm [12 in.])deep has been used to document storm events in sandy sediments (Beavers et al., 1997), but can also be used in muddy to gravelly sand sediment types. It has a removable slide hammer and angled sliding door, and is easily deployed by one person. A lift bag can be used if multiple samples are taken during one dive.

A diver-operated hydraulic underwater coring device has been used to take 4 cm (1.5 in) diameter cores in 1.5 m (4.5 ft.) increments, up to a coring depth of 20 m (60 ft.) (Macintyre, 1975; Shinn et al., 1990). It can also be operated in a handheld mode with larger diameter (10 cm/4 in.) diamond-tipped pipe to sample large coral heads. It consists of an underwater hydraulic wrench (Stanley Hydraulic Tools; Model IW23) fitted with drilling pipes and barrels, which is powered from the surface (Macintyre, 1996). The drilling unit can be very heavy, and is mounted on a tripod for vertical profiles, and when large diameter or long drilling pipes are used. A three-person crew is required for maximum drilling efficiency.

Geological samples of fossil coral skeletons have been collected by pounding galvanized steel pipe (2.5 cm/1 in.) diameter) into coral formations (Littler et al., 1996). Core increments of 20 cm/8 in. were taken from 6 levels, down to a depth of 6 m (20 ft.).

Rock core collections have also been conducted in cave environments, using a rebuilt Clarke air model pneumatic wrench (Schwabe, 2008). A diffuser unit was attached to the exhaust of the air drill to minimize the impact of gas bubbles on the cave ceiling.

Walker and Gurney (1985) describe a diver manipulated gravel suction pump used to recover diamonds from the surf zone off South Africa. The divers use a hookah air supply, and position 15 cm (6 in) diameter pipes which are connected on shore to tractor-mounted pumps. Up to 30 cubic meters of gravel and sediment are pumped each day.

Fallout from explosive submarine volcanic eruptions has been studied by scientific divers (Fiske and Avery, 1996). Fallout such as pumice and lithics have specific fallout rates through the water column, which were measured in a large (35 cm diameter, 3.4 m height) water-filled separation tube that was deployed from a small boat. At the bottom end, a drawer-like device with removable trays collected the falling particles. Various mixtures of particles are released at the top of the tube, and allowed to fall into the trays below.

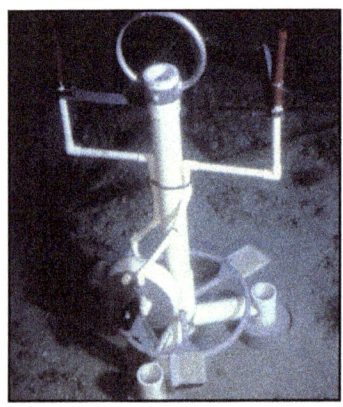

FIGURE 6.6:
Sediment traps can be placed by divers in specific locations.
(photo courtesy of J. Reed)

FIGURE 6.7:
Divers measure sand ripple wavelength and height.
(photo courtesy of J. Heine)

To measure and map sand ripples, divers can measure the wavelength and height of the ripples with a meter stick, measure their slope with an inclinometer, and their orientation with a compass. Using divers in a sandy near shore environment in California, Inman (1957) developed a relationship between wave orbitals and the size of sand ripples on the bottom. A diver-operated sonic profiler has also been used to measure wave-formed ripples (Dingler et al., 1977). For additional information on measuring sand ripples, see Inman (1957) and Newton (1968).

Measuring structural complexity (e.g., rugosity) on rocky bottoms or coral reefs is an important part of many studies. The standard method employed is chain-and-tape method, which generates a measure of rugosity based upon the ratio of contour-following vs. straight distance between two points (McCormick, 1994; Knudby and LeDrew, 2007). The importance of chain link size has been documented by comparing different sizes in substrate types such as sand, rubble, coarse branching coral, tabulate coral, foliose coral, and fine branching coral (Knudby and LeDrew, 2007).

FIGURE 6.8:
An inclinometer can be used to measure the slope of geological features.
(photo courtesy of J. Heine)

TRAINING EXERCISE #6
Geological Mapping

Equipment needed: compasses, meter sticks, inclinometer, core samplers, penetrometer, slates

Objective: To characterize the geological aspects of a site, including compass bearings, slope size and aspect, sediment size and characteristics

The class can be divided into buddy pairs or groups of three. A thorough briefing must be given by the teaching staff prior to the dive. A "dry run" on land is often beneficial prior to the dive. Each group will have at least one compass, one ruler, one inclinometer, one core sampler, and one slate to record data. Each group will be given an area to map underwater. Depending upon the local conditions, a suggested size might be 25 m by 25 m. It is desirable to have an area with a mixture of substratum types (i.e., solid rock, cobble, gravel, sand, and/or mud). One buddy group could lay out transect tapes to delineate the area. Another group can draw a rough map indicating depth, substratum type, and general compass bearings of sand ripples, if present. Other groups can use the penetrometer (if available) on a grid pattern to cover all types of substratum, measure sand ripples wavelength and height, measure sand ripple slope using the inclinometer compass, and take core samples of sediment on the crest and troughs of the ripples. The divers should rotate through the various tasks so that everyone has a chance to handle all of the equipment in the exercise. The data can be pooled in class for evaluation and write-up. A debriefing is helpful to discuss problems encountered and to solicit suggestions for improvement or modification of the method for particular circumstances or environments.

To measure maximum shear strength, Dill and Moore (1965) modified a torque screwdriver by adding a vane to the shaft. The vane was inserted into the sediment, and the torque was increased to the point of sediment failure, to obtain a measurement of maximum shear strength of surface sediments.

Fissures in a submarine canyon have been mapped and studied to determine if they were faults, joints, or erosional features (Webb, 1988). Strike measurements are recorded by holding a compass along a slate, which is inserted into the fissure. Dip measurements can be made by using a leveling device such as an inclinometer. A combination surveyors compass/inclinometer was put into an underwater housing, described by Shumway (1955).

Sediment traps and tubes are used for measuring the amount of sediment that accumulates over time. They can be constructed from a variety of materials such as PVC, steel, or aluminum, and a height to diameter ratio of at least 2:3 is recommended (Gardner, 1980; Butman, 1986).

FIGURE 6.9:
A diver operating a sediment pump in a seagrass meadow. (photo courtesy of A. Norro)

Groundwater discharge into various types of marine and freshwater habitats has been measured *in situ* by scientific divers using seepage meters and submarine piezometers (Simmons, 1986; Duncan et al., 2006). These devices are discussed in detail by Lee (1977) and Lee and Cherry (1978). They are pushed into the sediment, and have collection bags for measuring the volume and chemical constitution of groundwater collected. They can also be fitted with a manometer to measure hydraulic pressure in order to determine the vertical gradient of water movement (Duncan et al., 2006). They are considered a diver-intensive tool, and require multiple visits.

Geological investigations in submerged caverns and caves have only recently become possible, due to limitations in safely accessing these habitats by divers (http://www.safe-cavediving.com/NACDActionStatement.shtml).

FIGURE 6.10:
Water sampling from a geothermal spring.
(photo courtesy of G. Caramanna)

Physical Oceanography

SCIENTIFIC DIVERS SHOULD HAVE SOME FAMILIARITY with deploying and retrieving oceanographic instruments underwater. These instruments can range from small thermometers or flow meters to larger tidal gauges and buoys and moorings which might have current meters, light meters, CTDs, or other instruments attached to them. Large and/or expensive instruments often have pinger transponders on them to aid in relocation. A diver-held receiver gives the diver information on bearings and relative distance to the object (see RJE International).

FIGURE 6.11:
Divers installing a tide gauge.
(photo courtesy of A. Norro)

Mounting of oceanographic instruments will depend upon the location, substrate, depth, exposure to environmental factors such as currents, ice, waves and surge, and boat traffic. For soft substrate such as mud, silt, or sand, instruments can be attached to frames mounted to the bottom with sand anchors, which screw into the substrate. Also, pipe can be pounded into soft bottoms with a sledge hammer, or installed by pumping water through the pipe while pushing it into the sediment, and can be removed by using a diver-operated pipe puller (Spencer, 1989). For drilling and cementing into rock, see Chapter 4.

Temperature

MEASURING WATER TEMPERATURE can be as simple as reading from a hand-held thermometer, to downloading an electronic recording thermistor or CTD. Diver-held thermometers are usually encased in a stainless steel or plastic case to protect them from breakage. Thermographs that record onto film (Environmental Devices Corp., type 109) are accurate to within 0.2°C, and are easily deployed and retrieved by divers (Pitts, 1993). The time interval of measuring and recording temperature can be set prior to deployment.

Recent advances in technology have made it fairly easy and inexpensive to obtain long-term water temperature records at most SCUBA depths using solid state temperature data loggers. Models like Onset Corporation's Optic StowAway Temp are around 11 cm (4.5 in) long and less than 2 cm (1 in) in diameter, and easily fasten to the bottom, buoy lines, or other equipment. They can store over 32,000 data points with a ten-year battery life, and batteries can be replaced. The sampling cycle is set using a personal computer. Data can be downloaded directly into a computer after the logger is retrieved, or downloaded *in situ* with a small optical shuttle and then transferred to a computer. Other popular models include the HOBO Water Temp Pro v2, the TidbiT v2, and the Pendant Logger. VEMCO markets a Minilog-II-T waterproof data logger that has a 10-year battery life and can be used to depths up to 500 m (1650 ft.). Fouling organisms can affect data transfer, so users often wrap the part of the logger used for downloading with tape, and then remove the tape before downloading.

Conductivity, Temperature, and Depth (CTD)

SOME MODELS OF CTDs are compact enough for divers to swim with them underwater and collect discreet data at precise locations (e.g. SeaBird). Water samples can also be collected with plastic bottles for measuring salinity, nutrients, and other physical or chemical parameters, or with small Niskin bottles.

Bozanic (1993) describes an instrument that measures salinity, conductivity, dissolved oxygen, redox potential, temperature, and depth, that is mounted between double cylinders on the diver's back.

FIGURE 6.12:
HOBO temperature data logger.
(photo courtesy of Onset)

FIGURE 6.13:
A diver uses a multi-probe on a transect to measure pH, dissolved oxygen, and temperature. (photo courtesy of G. Caramanna)

Water Motion

WATER MOTION CAN BE A COMPLEX AND DIFFICULT parameter to measure (see Denny, 1985 for a review). A crude, relative measurement of total water motion has been done using plaster of paris spheres (Muus, 1968) or clod-cards (Doty, 1971). Blocks of plaster of paris are weighed, affixed to some type of framework that can be deployed in the field, and deployed for some period of time. The cards can then be collected, dried, and weighed to give a relative measure of water motion by difference in weight (Muus, 1968; Doty and Doty, 1973; Foster et al., 1985; Goldberg and Foster, 2002). A variation on this method assays for bulk fluid transport close to suspension feeding organisms (Genovese and Witman, 1997). Dissolution of rectangular alabaster slabs ($CaSO_4 \cdot 2H_2O$), which were spray painted with waterproof enamel on all but the top side, was measured by weight difference and surface area digitization. To relate relative dissolution rates to actual currents, alabaster slabs were also attached near an S4 current meter (see below). Airoldi and Cinelli (1997) used the rate of dissolution of gypsum balls (2.3 mm diameter) as a measure of flow microenvironment, using the preparation described by Gambi et al., (1989). Harger (1970) measured cumulative wave forces with a sliding metal plate on a nail. Water velocity perpendicular to the plate would force the plate down the nail relative to the strength of the force.

FIGURE 6.14:
Plaster clod card affixed to rocky substrate.
(photo courtesy of M. Edwards)

Another method of measuring current direction and velocity is by using fluorescent dye, which can be released by divers and visually tracked and timed, or recorded on a video camera with a timer (Maney et al., 1990). Denny (1983) describes a Teflon slider device that measures the direction and magnitude of a current or wave force. Surge can be measured using a modified protractor and a buoyant tethered sphere (Foster et al., 1985), which requires direct observation by a diver. In rough wave conditions, a remote drag sphere was used to measure turbulent water velocities (Donelan and Motycka, 1978). It used strain gauges to measure forces in three-dimensions. Jones and Demetropoulos (1968) measured the maximum force or velocity using a standard spring scale with a maximum recording lever.

FIGURE 6.15:
Use of dye to trace the water circulation around an underwater volcanic gas vent.
(photo courtesy of G. Caramanna)

Current Meters

CURRENT METERS HAVE been deployed in a variety of habitats, often by divers, especially in shallow water. Small, hand-held flow meters are available with different rotor sizes for different water velocity ranges (e.g., General Oceanics). They can be calibrated in a flume, or flow tank with a known current speed. Other types of fabricated flow probes have been reported in the literature (LaBarbera and Vogel, 1976). Denny (1985) describes the use of a simple commercial sailboat paddle wheel used to measure velocity.

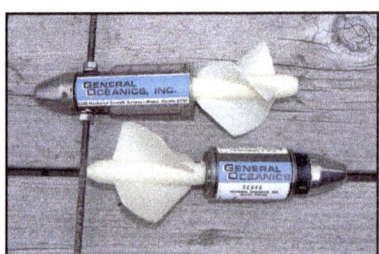

FIGURE 6.16:
Small, hand-held digital flow meters.
(photo courtesy of J. Heine)

FIGURE 6.17:
An S4 current meter is installed on a concrete base.
(photo courtesy of D. Kesling)

A current meter can be attached to a taut-line mooring that is weighted, anchored, or fixed to a sand anchor in soft sediment. Under strong currents, however, the current meter will be deflected, and a rigid mooring may be preferable (Pitts, 1990). An inexpensive rigid design is described by Pitts (1990). It consists of a concrete block with four outriggers for stability, and a vertical pole with a swivel on top for the current meter.

A commonly-used electromagnetic current meter (Model S4, Interocean Systems, Inc.) has been used extensively in near shore waters to measure direction and magnitude of currents. They are programmable, but very expensive.

Acoustic Doppler Current Profilers (ADCPs) record water velocities using a measured Doppler shift. They can be used in all depth ranges, and often include acoustic pingers for relocation (Work et al., 2006). For relocation, pop-up buoys can be triggered by an acoustic signal from a deck unit, which triggers a release of a tethered float that comes to the surface.

FIGURE 6.18:
A current meter with diver-operated lift bag for proper vertical positioning.
(photo courtesy of M. Abbiati)

Wave Meters

WAVE METERS, BY NECESSITY, ARE DEPLOYED in shallow water, often in the surf zone. Traditional anchor-type moorings tend to sink into the sand, sometimes burying the instrument. In this case, the instrument can be mounted on a pole, which can be driven into the sand using a water-injection pump (Holman et al., 2006; England et al., 2008).

Work et al. (2006) describe a wave buoy that contains three accelerometers, three angle rate sensors, a microcomputer, and four large batteries. Solar panels on the surface buoy are used to recharge the batteries. The Iridium satellite communications package sends the data to a remote computer.

FIGURE 6.19:
Wave meter deployed in a sandy bottom environment.
(photo courtesy of G. Symonds)

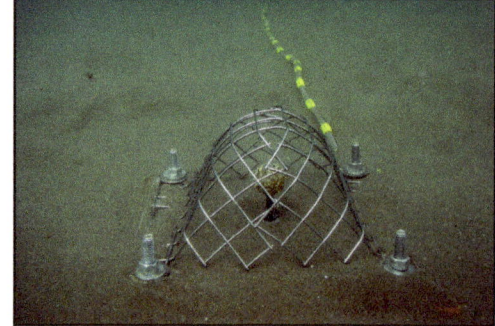

FIGURE 6.20:
Custom-made wave energy sensor to test the effects of cages on water flow.
(photo courtesy of C. Nelson)

Light and Color

MEASUREMENTS OF LIGHT ARE IMPORTANT for a variety of studies, including plant productivity measurements and physical oceanography projects (Ramus, 1985; Mobley, 1994). There are a number of commercially available light meters (irradiance, etc.) that measure photosynthetically active radiation (PAR, 400 to 700 nm wavelength) (see Refs.). Dunton et al. (1992) describe an LI-193SA spherical (4pi) quantum sensor and an LI-1000 Datalogger (Li-Cor, Inc.) housed in a watertight case, which were deployed on the seafloor in the Arctic for a one-year period. Instantaneous PAR was measured at one-minute intervals and integrated over a three-hour period. Hanisak (1997) compared arrays of various quantum sensors (collectors) over a one-year period. Fouling of the sensors was a significant problem, even with frequent cleaning. Goldberg and Foster (2002) used a hand-held cosine photon collector attached to a Li-Cor model 1400 data logger to measure irradiance in the horizontal and vertical planes in an experiment to test settlement of an alga in a kelp forest.

A diver-operated irradiance meter (Biospherical Instruments, Inc., model QSI-140), which measures PAR from 400 to 700 nm, has been used in a variety of studies (Richardson and Carlson, 1993). Gittings et al. (1990) describe the use of a Biospherical Instruments QSP 200L Submersible Quantum Scaler Irradiance meter. A custom-made radiometric instrument measures vertical profiles within and beneath vegetation canopies (HOBI Labs, Pand and Zimmerman, 2010).

FIGURE 6.21:
Diver-installed PAR sensor mounted on rebar in a kelp forest. (photo courtesy of C. Nelson)

FIGURE 6.22:
Diver-Operated Benthic Bio-Optical Spectrometer (DOBBS) in a kelp forest.
(photo courtesy of C. Nelson)

FIGURE 6.23:
Light meter, data logger, and spherical sensor in an Antarctic algal bed.
(photo courtesy of J. Heine)

FIGURE 6.24:
A scientific diver installing a light source and battery pack for optical instrument calibrations. (photo courtesy of J. Heine)

Fluorescence

THERE ARE A VARIETY OF DIVER-OPERATED DEVICES that measure small-scale fluorescence of organisms, and the optical properties of the sea floor and water above. Traditional methods include deploying large instruments or moorings from ships. Oceanographers can now place smaller instruments in specific locations or near particular organisms (Ackleson, 1996). A diver-operated instrument that measures the fluorescence excitation-emission spectrum of corals is described by Mazel (1995 and 1997) and Ackleson (1996). The electronics (Ocean Optics model S1000 Spectrometer, and more recently DiveSpec) are housed in Lexan containers, and an optical fiber connects the hand-held probe to the spectrometer. These instruments have been used for benthic mapping in shallow water (Mazel, 1997). The scientist can view the data underwater, and can decide to store the data or change the instrument settings. Stored data from these units are downloaded to a personal computer for processing.

FIGURE 6.25:
Bottom reflectance measured with a DiveSpec underwater spectrometer.
(photo courtesy of R. Nichols)

CHAPTER 6 MEASURING PHYSICAL FACTORS 111

FIGURE 6.26:
Diver-operated PAM fluorometer.
(photo courtesy of D. Andersen)

FIGURE 6.27:
In situ measurement of seagrass photosynthetic activity using PAM fluorometer.
(photo courtesy of D. Pence)

Chlorophyll *a*, the photosynthetic pigment common to all algal cells, can be measured with a fluorometer by a diver in the field (Dunton et al., 1992; Jones and Dennison, 1997; Eklund et al., 2008). The diving version of the mini-PAM fluorometer (Walz) offers a rapid and straightforward measurement of the state of photosynthesis in plants and algae *in vivo* (http://www.walz.com/products/chl_p700/diving-pam/introduction.html). Chlorophyll fluorescence is an indication of quantum yield of photosynthesis; and, therefore, fluorescence offers a quantitative measurement of environmental stress. PAM (Pulse-Amplitude-Modulation) fluorometers are suited for measuring fluorescence yield of dark adapted samples as well as samples in ambient daylight, and a portable mini-PAM was developed for terrestrial field work. The mini–PAM instrument requires little set-up, no sample preparations are necessary, and the measurement is rapid and non-intrusive to the plant. With the new diving–PAM, chlorophyll fluorescence can be assessed underwater or under more extreme weather conditions. Marine biologists have recently applied PAM technology to address impacts of UV light, temperature, and nutrient stresses on seagrasses and algae (see also Ch. 7). On recent Arctic and Antarctic expeditions, Dunton et al. (pers. comm.) have successfully utilized the dive–PAM in 30 m of water and under 2 m of ice. Other measures of photosynthesis, such as O_2 evolution or carbon isotope fixation, usually complement fluorescence yield experiments.

Another device for measuring inherent optical properties (IOP), is the ac-9 (absorption and beam attenuation measured at 9 wavelengths), which is manufactured by Wet Labs, Inc. (Ackleson, 1996). This flow-through instrument has a light source at one end and a detector at the other end of a 25 cm (10 in.) path. For a complete description of the instrument, see Moore et al. (1992). Other instruments are now in the testing phase that will mount in a double-cylinder backpack in place of one of the scuba cylinders.

A secchi disk or beam transmissometer can be used to measure the transparency or clarity of the water. A diver can swim a secchi disk horizontally until it just disappears from the view of another diver, who notes the distance from a marked line. Transmissometers are also useful as a measure of the concentration of suspended matter, but are generally not portable diver-carried devices (Ramus, 1985).

Sound

UNDERWATER ACOUSTIC MEASUREMENTS OFTEN require large transducers and hydrophones that can be deployed by divers. Piper (1990) describes an array of three transducers on a tripod that weighs 230 kg (500 lb) in water. Divers used lift bags to move and place the unit on the seafloor.

Hydrophone arrays are also used underwater to record sounds from both physical and biological sources. One array was designed to monitor and track harbor seals, and correlate recorded sounds with observed behaviors (Mellinger, 1997). The hydrophones are protected by a wire mesh cage to discourage damage by marine mammals.

Many fishes produce species-specific sound patterns that relate to courtship and mating behaviors (Lobel, 2001; Luczkovich and Sprague, 2003). These sounds can be measured with hydrophones. A unique design is shown by Lobel (2001), where he mounted a hydrophone on an extension pole from a video camera. This allowed the simultaneous recording of sounds and behavior of the fishes. He also used a rebreather to minimize noise from exhaled bubbles.

Lobel (2005) found that scuba bubbles produce noise in the same range as the hearing sensitivity in many fish spe-

FIGURE 6.28:
Diver installing a hydrophone by anchoring the cable to the sand with a screw-in earth anchor.
(photo courtesy of J. Murray)

CHAPTER 6 MEASURING PHYSICAL FACTORS 113

cies. Additionally, the use of relatively quiet, bubble-free rebreathers alleviates a significant source of disturbance to the fishes being observed, allowing for more accurate behavioral observations to be made.

Recordings of fish sounds can be played back underwater to study fish behavior and ecology (Luczkovich and Sprague, 2003). The recordings are made with hydrophones suspended from vessels, autonomous sonobuoys, and either diver-deployed or ROV underwater video. A fish sound catalog is being developed for use in the identification of unknown sounds.

Figure 6.29:
Hydrophone and video camera with quiet rebreather used to study fish bioacoustics and behavior. (photo courtesy of L. Lobel)

Fishes and marine mammals also make sounds for communication with conspecifics, and to signal alarm. Longspine squirrelfish produce low frequency vocalizations that are related to territorial behavior (Luczkovich and Keusenkothen, 2007). Playback sound studies using a Clark AC339 underwater speaker driven by a car amplifier were undertaken with divers using rebreathers while videotaping the behaviors.

Inexpensive point-and-shoot digital cameras in underwater housings have also been used to record underwater sounds (Kovitvongsa and Lobel, 2009). Various camera models were tested underwater with microphones facing a speaker that played sounds that varied in frequency, amplitude, and temporal patterns. Many of the systems produced inherent noise from their electronics that affected the recordings. The underwater housings also reduced the sound reaching the camera microphone by over one-half. Kovitvongsa and Lobel (2009) recommend that researchers use hydrophones for comprehensive studies of underwater acoustics.

TRAINING EXERCISE #7
Measurement of Physical Oceanography Parameters

Equipment needed: As available, thermometers, flow meters, current meters, light meters, CTD, underwater dye, water bottle sample collections, pH meter, hydrophone
Objective: To characterize the physical oceanographic aspects of a site

The class can be divided into buddy pairs or groups of three. A thorough briefing must be given by the teaching staff prior to the dive. A "dry run" on land is often beneficial prior to the dive. Equipment can often be borrowed from in-house research groups. Various scientific equipment can be set up in stations underwater, where the divers can rotate through the various tasks so that everyone has a chance to handle all of the equipment in the exercise. One variation on this exercise would be to collect data in a variety of habitats or locations, such as up current and down current of a reef or kelp forest, on the bottom, in mid water, and at the surface, or in an estuary, open ocean (blue water), or near shore environment. The data can be pooled in class for evaluation and write-up. A debriefing is helpful to discuss problems encountered and to solicit suggestions for improvement or modification of the method for particular circumstances or environments.

Large Buoys or Moorings

DIVING IS OFTEN USED FOR DEPLOYMENT, RETRIEVAL, monitoring, evaluation, and photographing large oceanographic instruments (Flemming and Max, 1996). Researchers at Woods Hole Oceanographic Institution describe how diving was used in the development phase of a mooring that uses variable buoyancy to travel to the surface once per day to transmit data to a satellite (Kery, 1989).

Moss Landing Marine Laboratories physical oceanography group is working with scientists from NOAA and NASA on a multi-year project to design, build and operate a sophisticated marine optical buoy called MOBY. The 16 m (50 ft.) long, one-ton MOBY measures light leaving the sea surface. One of MOBY's purposes is to compare light measurements near the sea surface with satellite measurements made 700 km overhead. MOBY will be used to confirm the accuracy of the SeaWiFS satellite measurements, and to obtain a long-term detailed time series of light in the sea.

Scientific divers are used to deploy and retrieve the MOBY buoys from a large research

Figure 6.30:
Divers working on the large MOBY buoy in the open ocean off Hawaii.
(photo courtesy of J. Heine)

vessel at a site 10 miles offshore the Hawaiian Islands. The divers handle lines, secure the buoy to a mooring, and take periodic calibration measurements of the light sensors. Working with hand tools around these types of large instruments in the open ocean from research vessels requires considerable planning, training, and careful execution.

Scientists at the Skidaway Institute of Oceanography have a number of coastal engineering and physical oceanographic projects on moorings, towers, and offshore platforms (Work et al., 2006). Divers install and service *in situ* sensors that measure currents, waves, and other parameters.

Chemical Measurements

SAMPLES OF SURFACE water, groundwater, and sediments are often required to evaluate contaminated or polluted sites, or for routine monitoring. These can be made remotely, but often require the use of divers for precise collections.

FIGURE 6.31:
Working in mid-water on oceanographic instruments requires excellent buoyancy control and handling of tools and equipment.
(photo courtesy of C. Storlazzi)

Chemical measurements by divers can range from simply collecting a water sample in a bottle or plastic bag, to using sophisticated collection and analyzing devices such as piezometers and diffusion samplers. Divers have mounted van Dorn bottles to their scuba cylinders, and tripped them at the appropriate time and place while underwater (Harper et al., 1989). For sample collection by hand, samples should be collected in polyethylene bottles that have been rinsed twice, which is best done by squeezing the bottles underwater to flush them out.

Nutrient fluxes (such as NO_3^-, NO_2^-, NH_4^+, PO_4^{-3}) in benthic chambers can be measured by withdrawing water samples by syringe and freezing for later analysis (Cahoon, 1986). Nutrient levels have been manipulated in the field by putting fertilizers in porous containers for slow release. Osmocote fertilizer (Sierra Chemical Co.) comes in pellet form, and has been released in mesh bags and trays (Foster et al., 1985).

Nitrogen is an important nutrient for plant growth, and has received the greatest attention, especially in the kelp forests of California (Zimmerman and Kremer, 1984). In times of El Niño conditions, high water temperatures generally have low nutrient levels, which affect kelp growth.

FIGURE 6.32:
Diver with PVC pipes and valves for water sample collection.
(photo courtesy of B. Robertson)

FIGURE 6.33:
Measuring dissolved oxygen in the sediment.
(photo courtesy of A. Norro)

Dissolved oxygen has been measured *in situ* by divers in caves using Chemetrix indicator ampoules (Bozanic, 1993). Powers et al. (1995) describe a dissolved oxygen and temperature logger (Endecco 1184) that was deployed by divers, which provided a continuous record for a three month period. A more sophisticated method employs the use of oxygen microelectrodes, which were employed to measure oxygen microprofiles *in situ* in the black band diseased (bacterial) areas of corals (Richardson and Carlton, 1993). They modified existing probes by using a thicker tip (ca. 150 µm) to reduce breakage (see Carlton and Wetzel, 1987). Accurate positioning of the microprobe was accomplished with a micromanipulator. These researchers used a modified microscope stage micrometer, which was less expensive and easier to maintain than a laboratory micrometer.

Divers have collected water samples in caves using silicon tubing and a three-way stopcock (Schwabe, 2008). The tubing was pre-filled with sterile water, then flushed with water from the cave, and collected in intravenous bags. Smaller volumes of water were collected in 10 ml vacutainers for bacterial counts.

Piezometers are hollow tubes that are inserted into the sediment for the collection of groundwater (Lee and Cherry, 1978; Duncan et al., 2006). They must be installed into a coarse enough sediment size to allow water movement. Diffusion samplers collect contaminants from the environment (pore water) via passive transport. They can be constructed from glass vials and membranes and be left in place for a period of days to weeks. They are often used in conjuction with piezometers to estimate averaged exposure point concentrations. Duncan et al. (2006) give a good review of methods and design recommendations for many of these techniques.

Diver-held pH probes have been used to document the level of kiln dust lechate discharge in Lake Michigan (Lee, 1990). Multi-parameter probes (e.g. In Situ Troll 9000 Professional multi-sensor) can be precisely placed by divers, with hard-wire connection to the surface. Parameters measured can include pH, conductivity, depth, temperature, oxidation-reduction potential, and dissolved oxygen (Duncan et al., 2006). A multi-probe data logger (Hydrolab DS3), which measured temperature, conductivity/salinity, and dissolved oxygen, was used in a cave environment in Mexico to obtain ceiling-to-floor vertical profiles (Talamante et al., 2008). The unit was mounted between manifold dive cylinders and also as a detachable unit from the diver's harness.

FIGURE 6.34:
Measuring sediment pH in a seagrass meadow.
(photo courtesy of A. Norro)

FIGURE 6.35:
A Unisense underwater micro-electrode system for measuring *in situ* DO and pH.
(photo courtesy of D. Andersen)

Geothermal active zones with gas vents (mainly CO_2) have been studied by divers in Italy (Caramanna et al., 2005). Gas samples were collected using an upside down plastic funnel with a Pyrex glass flask connected through a silicone hose. Divers used full-face masks and dry suits to avoid contact with the acidic pH 5 water.

FIGURE 6.36:
A diver inspects a large device for measuring the effects of advection.
(photo courtesy of A. Norro)

FIGURE 6.37:
Diver holding instruments for gas-flow measurements.
(photo courtesy of G. Caramanna)

Equipment Manufacturers and Suppliers

Aanderaa Instruments (wave and tide recorder)
182 East Street, Suite B
Attleboro, MA 02703-4209
(508) 226-9300
http://www.aadi.no/Default.htm

Benthos (oceanographic instruments, hydrophones, ROV)
49 Edgerton Dr.
No. Falmouth, MA 02556-2826
(508) 563-1000
Email: info@benthos.com
www.benthos.com

Biospherical Instruments, Inc. (light meters)
4901 Morena Blvd., #1003
San Diego, CA 92117
(619) 686-1888
http://www.biospherical.com/

Coastal Leasing, Inc. (wave gauges)
179 Sidney Street
Cambridge, MA 02139
(617) 497-1122
http://www.coastal-usa.com/

Falmouth Scientific, Inc. (acoustic sensors for environmental data acquisition)
1140 Route 28A
Cataumet, MA 02534
(508) 564-7640
www.falmouth.com

General Oceanics, Inc. (oceanographic instruments, hand-held flow meter)
1295 NW 163 Street
Miami, FL 33169
(305) 621-2882
http://www.generaloceanics.com/

Helle Engineering (pingers)
Julian, CA 92036

InterOcean Systems, Inc. (current meters, wave gauges)
3738 Ruffin Rd.
San Diego, CA 92123
(858) 565-8400
http://www.interoceansystems.com/

Klein Associates (side-scan sonar)
11 Klein Drive
Salem, NH 03079-1249
(603) 893-6131
http://www.l-3klein.com/

LI-COR, Inc. (light meters)
4547 Superior St.
Lincoln, NB 68504
(402) 467-3576
http://www.licor.com/

Onset Computer Corporation (temperature loggers)
P.O. Box 3450
Pocasset, Ma. 02559-9911
(800) LOGGERS
www.onsetcomp.com

RJE International (sonar, navigation, pingers, and receivers)
15375 Barranca Parkway, Suite B-107
Irvine, CA 92618
(949) 727-9399
www.rjeint.com/

Sea-Bird Electronics (oceanographic instruments)
13431 NE 20th Street
Bellevue, WA 98005
(425) 643-9866
http://www.seabird.com/

SonTek/YSI (oceanographic instruments)
9940 Summers Ridge Road
San Diego, CA 92121
(858) 546-8327
Email: inquiry@sontek.com
www.sontek.com

Subsalve USA (lift bags)
P.O. Box 2030
North Kingstown, RI 029852
(800) 466-6962
www.subsalve.com

VEMCO (temperature and depth data loggers, pingers, telemetry)
211 Horseshoe Lake Drive
Halifax, Nova Scotia
Canada
(902) 450-1700
www.vemco.com

Walz (mini-PAM)
Eichenring 6, 91090 Effeltrich
Germany
49 (0) 9133-7765-0
http://www.walz.com/products/chl_p700/mini-pam/introduction.html

Wet Labs, Inc. (fluorometers, transmissometers)
P.O. Box 518
620 Applegate St.
Philomath, OR 97370
(541) 929-5650
http://www.wetlabs.com/

YSI (water quality instrumentation)
1700/1725 Brannum Lane
Yellow Springs, OH 45387-1107
(800) 765-4974
Email: info@YSI.com
http://www.ysi.com/index.php

References

Ackleson, S.G. 1996. Diver-operated instruments to measure coastal benthic optical properties. In: Lang, M.A. and C.C. Baldwin (eds.), Methods and Techniques of Underwater Science, Proceedings of Amer. Acad. Underwater Sci. pp. 1-9.

Airoldi, L. and F. Cinelli. 1997. Effects of sedimentation on subtidal macroalgal assemblages: An experimental study from a Mediterranean rocky shore. J. Exp. Mar. Biol. Ecol. 215: 269-288.

Akal, T. 1985. The use of diving techniques for *in situ* geoacoustic measurements on the sea-floor. In: Mitchell, C.T. (ed.), Diving for Science...85, Proceedings of Joint International Scientific Diving Symposium, Amer. Acad. Underwater Sci. pp. 273-281.

Anuskiewicz, R.J. and E.G. Garrison. 1992. Underwater archaeology by Braille: Survey methodology and site characterization modeling in a blackwater environment - a study of a scuttled confederate ironclad, *C.S.S. Georgia*. In: Cahoon, L.B. (ed.), Diving for Science...1992, Proceedings of Amer. Acad. Underwater Sci. pp. 1-12.

Anuskiewicz, R.J. and J.S. Dunbar. 1993. Evidence of prehistoric man at Ray Hole Springs: A drowned sinkhole located 32 km offshore on the continental shelf in 12 m seawater. In: Heine, J.N. and N.L. Crane, (eds.), Diving for Science...1993, Proceedings of Amer. Acad. Underwater Sci. pp. 1-11.

Beavers, R., T. Boynton, and P. Howd. 1997. Documenting storm sedimentation with diver-collected box cores and acoustic altimeters offshore of Duck, NC. In: Maney, Jr. E.J. and C.H. Ellis, Jr. (eds.), Diving for Science...1997, Proceedings of Amer. Acad. Underwater Sci. pp. 1-4.

Bozanic, J. 1993. Preliminary investigations in anchialine caves of Cuba. In: Heine, J.N. and N.L. Crane, (eds.), Diving for Science...1993, Proceedings of Amer. Acad. Underwater Sci. pp. 33-41.

Breiner, S. and R.G. MacNaughton. 1965. The application of magnetometers to underwater archaeology. Second Conference on Underwater Archaeology, Toronto.

Butman, C.A. 1986. Sediment trap biases in turbulent flow: Results from a laboratory flume study. J. Mar. Res. 44: 645-693.

Cahoon, L.B. 1986. The role of sediment-water column interactions in the continental shelf ecosystem. In: Mitchell, C.T. (ed.), Diving for Science...86, Proceedings of Amer. Acad. Underwater Sci. pp. 171-180.

Caramanna, G., N. Voltattorni, L. Caramanna, D. Cinti, G. Galli, L. Pizzino, and F. Quattrocchi. 2005. Scientific diving techniques applied to the geomorphological and geochemical study of some submarine volcanic gas vents (Aeolian Islands - Southern Tyrrhenian Sea - Italy). In: Godfrey, J.M. and S.E. Shumway (eds.), Diving for Science 2005, Proceedings of Amer. Acad. Underwater Sci. pp. 217-228.

Carlton, R.G. and R.G. Wetzel. 1987. Distributions and fates of oxygen in periphyton communities. Can. J. Bot. 65: 1031-1037.

Cood, D. and D. Gorsline. 1972. Field observations of sand transport by shoaling waves. Mar. Geol. 13: 31-56.

Coyer, J., and J. Witman. 1990. The Underwater Catalog: A Guide to Methods in Underwater Research. Shoals Marine Laboratory, Cornell University, NY.

Coyer, J., D. Steller, and J. Witman. 1999. The Underwater Catalog: A Guide to Methods in Underwater Research. 2nd ed. Shoals Marine Laboratory, Cornell University, Ithaca, NY. 150 pp.

Denny, M.W. 1983. A simple device for recording the maximum force exerted on intertidal organisms. Limnol. Oceanogr. 28: 1269-1274.

Denny, M.W. 1985. Water Motion, pp. 7-32. In: Littler, M.M. and D.S. Littler (eds.). Handbook of Phycological Methods. Ecological Field Methods: Macroalgae. Cambridge University Press. 617 pp.

Dill, R. and D. Moore. 1965. A diver-held vane-shear apparatus. Mar. Geol. 3: 323-327.

Dill, R. and G. Shumway. 1954. Geological use of self-contained diving apparatus. Bull. Am. Assoc. Petroleum Geol. 38(1): 148-157.

Dingler, J.R., J.C. Boylls, and R.L. Lowe. 1977. A high-frequency sonar for profiling small-scale subaqueous bedforms. Mar. Geol. 24: 279-288.

Dingler, J.R. 1988. Advances in underwater geologic research on near shore sand-bed dynamics. In: M.A. Lang (ed.), Advances in Underwater Science...1988, Proceedings of Amer. Acad. Underwater Sci. pp. 27-31.

Donelan, M.A. and J. Motycka. 1978. Miniature drag sphere velocity probe. Rev. Sci. Instr. 49: 298-304.

Dornblaser, M.M., J. Tucker, G.T. Banat, K.H. Foreman, M.C. O'Brien and A.E. Giblin. 1989. Obtaining undisturbed sediment cores for biogeochemical process studies using scuba. In: Lang, M.A. and W.C. Jaap, (eds.), Diving for Science...1989, Proceedings of Amer. Acad. Underwater Sci. pp. 97-104.

Doty, J.E. and M.S. Doty. 1973. Abrasion in the measurement of water motion with the clod-card technique. Bull. So. Calif. Acad. Sci. 72:40-41.

Doty, M.S. 1971. Measurement of water movement in reference to benthic algal growth. Bot. Mar. 14: 32-35.

Drew, E.A., J.N. Lythgoe, and J.D. Woods (eds.). 1976. Underwater research. Academic Press, New York. 430 pp.

Duncan, P.B., R. Henry, R. Pederson, S. Sheldrake, and D. Thompson. 2006. Adaptation of groundwater evaluation and sampling tools for underwater deployment. In: Godfrey, J.M. and N.W. Pollock (eds.), Diving for Science, Proceedings of Amer. Acad. Underwater Sci. pp. 55-84.

Dunton, K.H., L.R. Martin, and G.S. Mueller. 1992. Seasonal and annual variations in the underwater light environment of an arctic kelp community. In: Cahoon, L.B. (ed.), Diving for Science...1992, Proceedings of Amer. Acad. Underwater Sci. pp. 83-92.

Eklund, N.G.A., C.A. Nygard, R. Nordstrom and A.M. Gylle. 2008. *In situ* study of relative electron transport rates in the marine macroalga *Fucus vesiculosus* in the Baltic Sea at different depths and times of year. Jour. Applied Phyc. 20: 751-756.

England, P.R., J. Phillips, J.R. Waring, G. Symonds, and R. Babcock. 2008. Modeling wave-induced disturbance in highly biodiverse marine macroalgal communities: Support for the intermediate disturbance hypothesis. Mar. and Freshw. Res. 59 (6): 515-520.

Fisher, R. and R. Mills. 1952. Sediment trap studies of sand movement in La Jolla Bay. Bull. Geol. Soc. Am. 63: 1328 (abstract).

Fiske, R.S. and V.F. Avery. 1996. Use of scuba in the study of explosive submarine volcanism. In: Lang, M.A. and C.C. Baldwin (eds.), Methods and Techniques of Underwater Science, Proceedings of Amer. Acad. Underwater Sci. pp. 121-126.

Flemming, N.C. and M.D. Max. 1996. Scientific Diving: A General Code of Practice. Best Publishing, 278 pp.

Foster, M.S., T.A. Dean, and L.E. Deysher. 1985. Subtidal Techniques, pp. 189-232. In: Littler, M.M. and D.S. Littler (eds.). Handbook of Phycological Methods. Ecological Field Methods: Macroalgae. Cambridge University Press. 617 pp.

Gambi, M.C., M.C. Buia, E. Casola, and M. Scardi. 1989. Estimates of water movement in *Posidonia oceanica* beds: A first approach. In: Boudouresque, C.F., A. Meinesz, E. Fresi, and V. Gravez (eds.), International Workshop on *Posidonia* beds, GIS Posidonie, France, vol. 2, pp. 101-112.

Gardner, W. 1980. Field assessment of sediment traps. J. Mar. Res. 38: 41-52.

Genovese, S.J. and J.D. Witman. 1997. Particle flux measurement in deep (25-40m) suspension-feeding communities in the Gulf of Maine. In: Maney, E.J., Jr., and C.H. Ellis, Jr. (eds.), Diving for Science...1997, Proceedings of Amer. Acad. Underwater Sci. pp. 59-69.

Gittings, S.R., K.J.P. Deslarzes, B.S. Holland, and G.S. Boland. 1990. Ecological monitoring on the Flower Garden Banks: Study design and field methods. In: W.C. Jaap, (ed.), Diving for Science...1990, Proceedings of Amer. Acad. Underwater Sci. pp. 107-118.

Goldberg, N.A. and M.S. Foster. 2002. Settlement and post-settlement processes limit the abundance of the geniculate coralline alga *Calliarthron* on subtidal walls. J. Exp. Mar. Biol. Ecol. 278: 31-45.

Hanisak, M.D. 1997. Continuous monitoring of underwater light in Indian River lagoon: Comparison of cosine and spherical sensors. In: Maney, Jr. E.J. and C.H. Ellis, Jr. (eds.), Diving for Science...1997, Proceedings of Amer. Acad. Underwater Sci. pp. 71-85.

Harger, J.R.E. 1970. The effect of wave impact on some aspects of sea mussels. Veliger 12: 401-414.

Harper, D.E., R.R. Salzer, L.D. McKinney, and J.M. Nance. 1989. Soft bottom benthos sampling using scuba. In: Lang, M.A. and W.C. Jaap, (eds.), Diving for Science...1989, Proceedings of Amer. Acad. Underwater Sci. pp. 145-152.

Holman, R.A., G. Symonds, E.B. Thornton, and R. Ranasinghe. 2006. Rip spacing and persistence on an embayed beach. Journ. Geophys. Res.-Oceans 111: C1.

Inman, D.L. 1957. Wave-generated ripples in near shore sands. Dept. of the Army, Corps of Engineers Tech. Memo 100. 65 pp.

Jones, W.E. and A. Demetropoulos. 1968. Exposure to wave action: Measurements of an important ecological parameter on rocky shores of Anglesey. J. Exp. Mar. Biol. Ecol. 2: 46-53.

Jones, A.B. and W.C. Dennison. 1997. Photosynthetic capacity of coral reef systems: Investigations into ecological application of the underwater PAM fluorometer. In: J.G. Greenwood and N.J. Hall (eds.), Proc. Australian Coral Reef Soc., Brisbane, Australia, pp. 105-118.

Kery, S.M. 1989. Diving in support of buoy engineering: The RTEAM project. In: Lang, M.A. and W.C. Jaap, (eds.), Diving for Science...1989, Proceedings of Amer. Acad. Underwater Sci. pp. 191-197.

Knudby, A. and E. LeDrew. 2007. Measuring structural complexity on coral reefs. In: Pollock, N.W. and J.M. Godfrey, (eds.), Diving for Science 2007, Proceedings of Amer. Acad. Underwater Sci. pp. 181-188.

Kovitvongsa, K.E. and P.S. Lobel. 2009. Convenient fish acoustic data collection in the digital age. In: N.W. Pollock (ed.), Diving for Science 2009, Proceedings of Amer. Acad. Underwater Sci. pp. 43-57.

LaBarbera, M. and S. Vogel. 1976. An inexpensive thermistor flowmeter for aquatic biology. Limnol. Oceanogr. 21: 750-756.

Lee, D. R. 1977. A device for measuring seepage flux in lakes and estuaries. Limnol. Oceanogr. 22: 140-147.

Lee, D.R. 2000. Delineation, quantification and mitigation of discharging plumes. In: [USEPA] Proceedings of the Ground-Water/Surface-Water Interactions Workshop. EPA 542/R-00/007, 2000: 35-38.

Lee, D. R. and J. A. Cherry. 1978. A field exercise on ground water flow using seepage meters and mini-piezometers. Jour. Geol. Ed. 27: 6-10.

Littler, M.M., D.S. Littler, B.L. Brooks, and J.F. Koven. 1996. Field-expedient ecological methods: A comparison between limited vs. optimal resources for the study of a unique coral reef. In: Lang, M.A. and C.C. Baldwin (eds.), Methods and Techniques of Underwater Science, Proceedings of Amer. Acad. Underwater Sci. pp. 159-165.

Lobel, P.S. 2001. Fish bioacoustics and behavior: Passive acoustic detection and the application of a closed-circuit rebreather for field study. Mar. Tech. Soc. Jour. 35 (2): 19-28.

Lobel, P.S. 2005. Scuba bubble noise and fish behavior: a rationale for silent diving technology. In: Godfrey, J.M. and S.E. Shumway (eds.), Diving for Science 2005, Proceedings of Amer. Acad. Underwater Sci. pp. 49-59.

Luczkovich, J.J. and M. Keusenkothen. 2007. Behavior and sound production by Longspine squirrelfish *Holocentrus rufus* during playback of predator and conspecific sounds. In: N.W. Pollock and Godfrey, J.M. (eds.), Diving for Science 2007, Proceedings of Amer. Acad. Underwater Sci. pp. 127-134.

Macintyre, I.G. 1975. A diver-operated hydraulic drill for coring submerged substrates. Atoll Research Bull. 185: 21-26.

Macintyre, I.G. 1996. A diver-operated submersible drill for studying the geological history of coral reefs. In: Lang, M.A. and C.C. Baldwin (eds.), Methods and Techniques of Underwater Science, Proceedings of Amer. Acad. Underwater Sci. pp. 167-174.

Maney, E.J., J. Ayers, K.P. Sebens, and J.D. Witman. 1990. Quantitative techniques for underwater video photography. In: W.C. Jaap, (ed.), Diving for Science…1990, Proc. of Amer. Acad. Underwater Sci. 1990. pp. 255-265.

Mazel, C.H. 1995. Spectral measurements of fluorescence emission in Caribbean coral reef cnidarians. Mar. Ecol. Prog. Ser. 120: 185-191.

Mazel, C.H. 1997. Diver-operated instrument for *in situ* measurement of spectral fluorescence and reflectance in benthic marine organisms and substrates. Opt. Eng. 36: 2612-2617.

McCormick, MI. 1994. Comparison of field methods for measuring surface-topography and their associations with a tropical reef fish assemblage. Mar. Ecol. Prog. Ser. 112(1-2): 87-96.

Mellinger, D.K. 1997. A low-cost, high-performance sound capture and archiving system for the subtidal zone. Proc. Inst. Acoustics 19(9): 115-122.

Merifield, P. and D. Rosencrantz. 1966. A simple method for surveying a small area underwater. Limnol. and Oceanogr. 11(3): 408-409.

Mobley, C.D. 1994. Light and water. Academic Press, San Diego, CA.

Moore, C., J.R.V. Zaneveld, and J.C. Kitchen. 1992. Preliminary results of an *in situ* spectral absorption meter. In: G.D. Gilbert (ed.), Ocean Optics XI, Proc. SPIE 1750: 330-337.

Muus, B.J. 1968. A field method for measuring "exposure" by means of plaster balls. Sarsia 34: 61-68.

Newton, R. 1968. Internal structure of wave-formed ripple marks in the near shore zone. Sedimentology 11: 275-292.

Pan, X., and R. Zimmerman. 2010. Modeling the vertical distributions of downwelling plane irradiance and diffuse attenuation coefficient in optically deep waters. J. Geophys. Res., 115, C08016, doi:10.1029/2009JC006039.

Piper, J.N. 1990. Diving in support of an underwater acoustic investigation in a very shallow water, near shore environment. In: W.C. Jaap, (ed.), Diving for Science…1990, Proceedings of Amer. Acad. Underwater Sci. pp. 293-299.

Pitts, P.A. 1990. A practical and effective mooring for current meters in shallow-water, high current speed locations. In: W.C. Jaap, (ed.), Diving for Science...1990, Proceedings of Amer. Acad. Underwater Sci. pp. 299-302.

Pitts, P.A. 1993. Coastal upwelling off the central Florida Atlantic coast: Cold near-shore waters during summer months surprise many divers. In: Heine, J.N. and N.L. Crane, (eds.), Diving for Science...1993, Proceedings of Amer. Acad. Underwater Sci. pp. 99-106.

Ramus, J. 1985. Light, pp. 33-52. In: Littler, M.M. and D.S. Littler (eds.). Handbook of Phycological Methods. Ecological Field Methods: Macroalgae. Cambridge University Press. 617 pp.

Richardson, L.L. and R.G. Carlton. 1993. Behavioral and chemical aspects of black band disease of corals: An *in situ* field and laboratory study. In: Heine, J.N. and N.L. Crane, (eds.), Diving for Science...1993, Proceedings of Amer. Acad. Underwater Sci. pp. 107-116.

Rule, N. 1989. The direct survey method of underwater survey, and its application underwater. The International Journal of Nautical Archaeology and Underwater Exploration. 18(2): 157-162.

Schwabe, S.J. 2008. The difficulties of sampling in underwater caves in the Bahamas: An exercise in ingenuity and survival. In: Brueggeman, P. and N.W. Pollock, (eds.), Diving for Science 2008, Proceedings of Amer. Acad. Underwater Sci. pp. 147-157.

Shinn, E.A., R.B. Halley, J.L. Kindinger, J.H. Hudson, and R.A. Slater. 1990. Underwater research methods for study of nuclear bomb craters, Enewetak, Marshall Islands. In: W.C. Jaap, (ed.), Diving for Science...1990, Proceedings of Amer. Acad. Underwater Sci. pp. 327-344.

Sigl, W., V. VonRad, H.J. Oeltzschner, K. Braune, and F. Fabricius. 1969. Diving sled: A tool to increase the efficiency of underwater mapping by scuba divers. Mar. Geol. 7:357-363.

Simmons, G. M. and J. Netherton. 1986. Groundwater discharge in a deep coral reef habitat: Evidence for a new biogeochemical cycle? In: Mitchell, C.T. (ed.), Diving for Science...86, Proceedings of Amer. Acad. Underwater Sci. pp. 1-12.

Spencer, W.D. 1988. Techniques for near shore oceanographic instrument installations. In: M.A. Lang (ed.), Advances in Underwater Science...1988, Proceedings of Amer. Acad. Underwater Sci. pp. 297-308.

Talamante, O.T., E. Escobar, P.A. Beddows, and J. Yager. 2008. Performing multiprobe profiles and hydrographic description of the water column in an anchialine cave in Quintana Roo. In: Brueggeman, P. and N.W. Pollock, (eds.), Diving for Science 2008, Proceedings of Amer. Acad. Underwater Sci. pp. 177-186.

Underwater Archaeology online:

http://library.thinkquest.org/27423/univer.htm

www.archaeology.org/wwwarky/underwater.html

http://nautarch.tamu.edu/

http://www.ecu.edu/cs-cas/maritime/

http://www.wisconsinhistory.org/archaeology/

Waddell, P.J.A. 1990. Electronic mapping of underwater sites. In: T.L. Carrell (ed.), Underwater Archaeology Proceedings from the Society for Historical Archaeology Conference. pp. 57-62.

Walker, C. H. and J.J. Gurney. 1985. The recovery of diamonds from the surf zone of the south Atlantic near the Olifants River, R.S.A. In: Mitchell, C.T. (ed.), Diving for Science...85, Proceedings of Joint International Scientific Diving Symposium, Amer. Acad. Underwater Sci. pp. 318-330.

Webb, D.A. 1988. A structural interpretation of Scripps submarine canyon. In: M.A. Lang (ed.), Advances in Underwater Science...1988, Proceedings of Amer. Acad. Underwater Sci. pp. 213-220.

Work, P.A., T. Moore, and K. Haas. 2006. Waves and currents in the Georgia Bight: Scientific diving in Georgia and South Carolina. In: Godfrey, J.M. and N.W. Pollock (eds.), Diving for Science, Proceedings of Amer. Acad. Underwater Sci. pp. 39-54.

Zimmerman, R.C. and J.N. Kremer. 1984. Episodic nutrient supply to a kelp forest ecosystem in southern California. J. Mar. Res. 42: 591-604.

Zimmerman, R. and A. Dekker. 2006. Aquatic optics: Basic concepts for understanding how light affects seagrasses and makes them measurable from space. In: A. Larkum, R. Orth, and C. Duarte (eds.) seagrasses: Biology, Ecology and Conservation. Springer, Dordrecht, Ch. 12, pp. 295-301.

CHAPTER 7

MEASURING BIOTIC FACTORS AND PROCESSES:
MATERIALS AND METHODS FOR UNDERWATER SAMPLING AND EXPERIMENTATION

Research Diving Tools

SCIENTIFIC DIVERS GENERALLY USE SIMPLE TOOLS to conduct biological research underwater. Many innovative techniques and materials have been invented by industrious researchers. Various methods of recording data, measuring distribution and abundance of benthic and pelagic organisms, types of quadrats, transects, collecting devices, tags, cages, transplantation techniques, methods for measuring recruitment, telemetry, and behavior will be discussed below.

Before selecting a tool or technique to use in underwater research, one must critically evaluate its utility for the objectives being proposed. This includes a search of the literature on the questions being asked. A number of relevant papers are listed in the reference section at the end of this chapter. In some cases, additional training in the use of these tools will also be required. Check with your diving safety officer or diving supervisor.

Data Collection and Recordkeeping

THE NEED TO RECORD DATA UNDERWATER IS REQUIRED for nearly every scientific project. This can be accomplished in a variety of ways. Perhaps the simplest is to write with a pencil directly on a slate constructed of styrene, or polyvinylchloride (PVC) (see Foster, 1976; Foster et al., 1985). Various types of pencils, including mechanical pencils, can be used. Short "golf" pencils are particularly good. Several extra pencils should be taken on every dive, as it is extremely difficult to sharpen pencils underwater. A good place to put them is in the ends of surgical tubing. Some people sharpen both ends of the pencil as well.

Slates can vary in size, and can even be curved to fit around the arm using surgical tubing or Velcro for attachment. Slate materials can be purchased for a lower cost in large sheets, and cut down to the desired size. A formatted data template can be drawn onto the slate with permanent ink. One disadvantage to this method is that the data must be transcribed later, either by hand or by photocopying. Data sheets of Mylar drafting film or other plastic sheets can be used as a more permanent record. Another popular method is the use of underwater paper (Nalgene Polypaper, Rite in the Rain), onto which formatted data sheets can be photocopied. Care must be taken, however, as this paper can melt under high temperatures found in some photocopiers. Data sheets can be secured onto the slate using surgical tubing bands, bungee cord, or a bolt-down type outer frame using wing nuts. Pencils should be connected to the slate via surgical tubing, string, or other appropriate cord. A fun class exercise can be arranged where each diver builds his or her own slate.

FIGURE 7.1:
Scientific diver recording information on a slate.
(photo courtesy of T. Wagner)

Some researchers have combined a variety of instruments onto one large board, to facilitate data collection (Dowling, 1963). These data boards might include things such as data sheets, a depth gauge, compass, protractor, inclinometer, ruler, and thermometer.

An underwater data recording tablet and portable computer graphical database system has been developed (Max and Nagel, 1997). The "E-map" consists of a Geographical Information System (GIS) navigational chart, with various assigned symbols for items such as graphical, text, numeric, and image information. It was originally developed for sea floor mapping purposes, such as sediment type, slope, penetrator tests, ripples, and common biota, but can be adapted for a wide variety of uses. Data are recorded onto Mylar overlays, and transferred to the database on the surface.

FIGURE 7.2:
Scientific divers use a variety of data sheets and slates.
(photo courtesy of L. Hesla/PISCO)

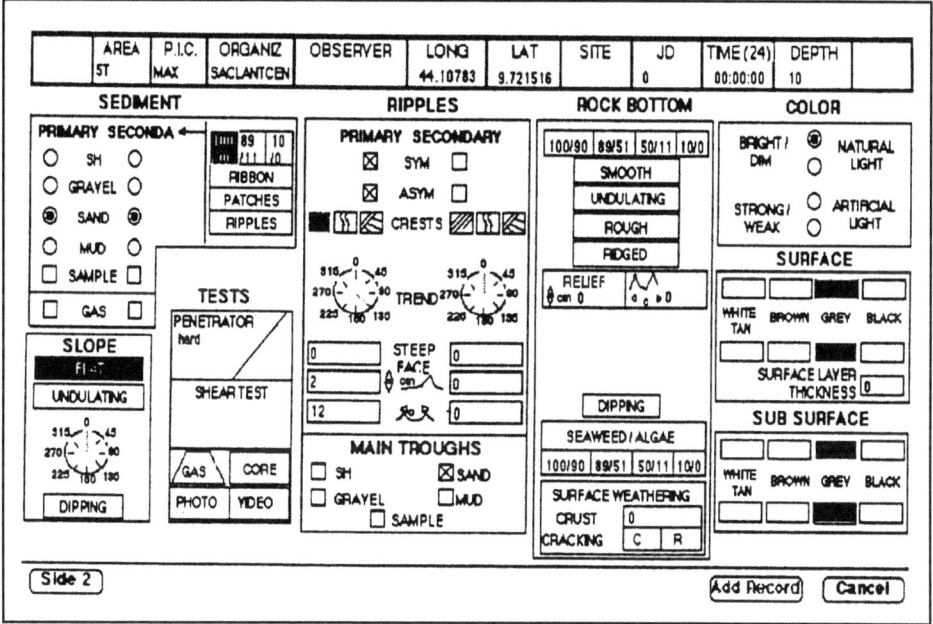

FIGURE 7.3:
Example of the general tablet of the "E-map."
(from Max and Nagel, 1997 AAUS Proc.)

Some studies require a large amount of data to be recorded, and might also require the diver to not be able to look down at a slate to write. In this case, recording voice data onto a tape or to a surface tender would be appropriate. Underwater tape recorders can be adapted to fit into waterproof cases (such as made by Ikelite), but require a modified regulator mouthpiece or full-face mask to be able to speak clearly underwater. A relatively inexpensive method uses the Buddy Phone, manufactured by Ocean Technology Systems, which is a wireless diver to diver, and diver(s) to surface communications system. A tape recorder can be plugged into the surface unit for recording data. Surface supplied diving, using a full face mask and umbilical line to the surface, can produce very clear communications, but can be expensive, and requires additional training and support personnel, as well as a large, stable platform to operate from (see Communications, Ch. 3).

Primary Productivity and Respiration Measurements

IN SITU MEASUREMENTS OF PRIMARY PRODUCTIVITY have many advantages over laboratory measurements, but physical parameters are more easily manipulated in the laboratory (Heine, 1983; Miller et al., 2009; see Littler and Arnold, 1985, for complete review). There are many different methods of measuring primary productivity, many of which can be carried out *in situ*. Freshwater algae have been incubated *in situ* in small 28 ml glass bottles containing ^{14}C bicarbonate (Campbell and Spence, 1976). Divers clipped the bottles onto a frame placed at the depth the plants were collected. The marine angiosperm *Posidonia oceanica*, which forms lush meadows in the Mediterranean, was incubated in plastic tubes (30 cm long, 1.5 to 2.5 cm internal diameter) using the ^{14}C method (Drew and Jupp, 1976).

FIGURE 7.4:
Light and dark bottles incubating macroalgae with magnetic stirring apparatus and air cylinder in the background.
(photo courtesy of J. Heine)

Marine algae have been enclosed in plastic bags with ^{14}C (Towle and Pearse, 1973), and in Bell jars with magnetic stir bars that were laid in Plexiglas trays and incubated underwater (Heine, 1983). The trays were periodically placed over air-driven magnetic stirrers to allow for water movement inside the jars, thus minimizing boundary layer problems and the potential for oxygen bubbles to form. Drew et al. (1976) describe the entire operation of conducting primary productivity experiments nearly entirely underwater, from collecting plants, to cutting disks of plant frond, to sealing the productivity jars, to injecting ^{14}C bicarbonate into the jars.

FIGURE 7.5:
Whole-plant productivity can be measured by enclosing the plant in a plastic bag.
(photo courtesy of M. Foster)

FIGURE 7.6:
Individual blades of kelp enclosed in polyethylene bag for measurement of exuded DOC. (photo courtesy of C. Nelson)

FIGURE 7.7:
Primary productivity chamber with spherical light collector and submersible pump. (photo courtesy of C. Nelson)

Other types of primary productivity or benthic respiration chambers, such as Plexiglas domes, fit right over the substrate, and have various ports that allow for the injection of radioactive tracers and removal of water samples for chemical analyses (Aloi et al., 1985; Cahoon, 1986). The chambers can be stirred with plastic rotors that are driven by local water motion. Automated benthic landers, which measure light flux and dissolved oxygen in replicate clear and opaque chambers, have been used in benthic habitats to measure primary production and respiration (Cahoon, 1996). For an excellent review of productivity measurements in open systems, see Kinsey (1985).

Miller et al., (2009) used the chamber depicted in Figure 7.7 to measure primary production and respiration of the benthos *in situ*. The chamber was constructed of clear rigid acrylic and fiberglass-reinforced plastic, and had an internal volume of 45 L. It was anchored to the bottom with a weighted flexible plastic skirt.

Repeated standing crop measurements have also been used as a measure of plant productivity. It is necessary to collect all of the plant biomass in a given area (e.g., 0.5 or 1 m^2), place it into bags, and transport to the surface for sorting, measuring, and weighing (Breda, 1982). Drew and Jupp (1976) describe the high standing crop of *Posidonia oceanica*, which forms lush meadows in the Mediterranean. Many studies of the standing crop of the giant kelp, *Macrocystis pyrifera*, have been conducted, showing a range of 0.7 to 22 wet kg/m^2 (North, 1971; Aleem, 1973; Gerard, 1976).

Organic matter from plankton blooms can fall to the bottom, and is considered to be a major source of energy in some habitats. Quantitative samples of the bottom "fluff" have been collected using a flume, as described by Schick et al., (1988). Water is drawn into a box by an electric pump and across a filter, which collects the fluff.

Macroalgal growth and productivity have been measured by punching holes in algal blades and measuring the movement of the holes over time (Mann, 1972). Other species have been stained with calcofluor white, which tags existing tissue with a fluorescent marker, which can be determined with a fluorescent microscope (Waaland and Waaland, 1975). Algal frond growth has also been determined *in situ* by comparing successive photographs taken over time (Johansen and Austin, 1970), or by simply measuring increases in plant length (Gerard, 1976).

FIGURE 7.8:
Benthic respiration chamber containing a sponge.
(photo courtesy of D. Gouge)

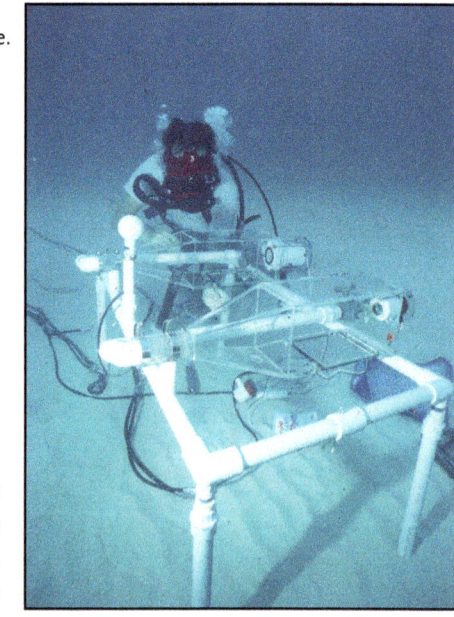

FIGURE 7.9:
Diver preparing benthic respiration chambers for an experiment.
(photo courtesy of D. Kesling)

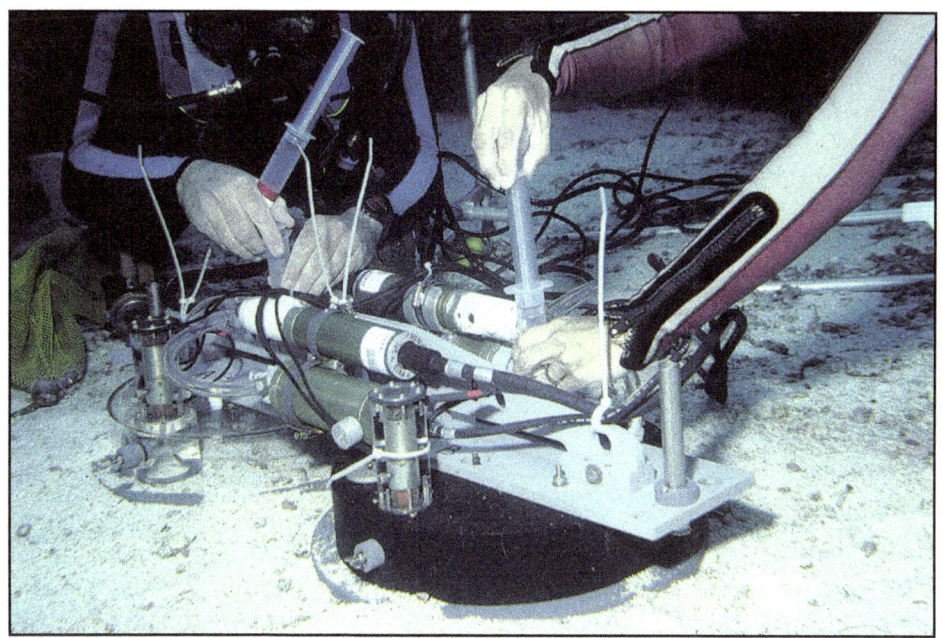

FIGURE 7.10:
Divers taking a water sample from a respiration chamber.
(photo courtesy of Q. Dokken)

Sediment oxygen demand can be measured using benthic flux chambers outfitted with oxygen electrodes and data loggers (YSI 600 series Sondes) (Balcom et al., 2007). The devices used in this study were dual-chambered, with each chamber having an internal volume of 13 L and covering 0.1 m^2 of surface sediment. Divers sampled water from the chambers using syringes. A review of the design and function of benthic chambers and profiling landers can be found in Tengberg et al., 1995.

Physiology

THERE HAVE BEEN A LARGE NUMBER OF PHYSIOLOGICAL experiments conducted *in situ* in many different environments with many different organisms. Sponges have been studied extensively (see Rutzler, 1978). Oscular currents and the microclimate around sponges have been measured with a thermistor flowmeter (Forstner and Rutzler, 1970). Respiration and metabolism chambers have been fabricated from plastic bell jars fitted with magnetic stirrers and electrode ports. Various types of underwater incubation devices have been used to study processes such as productivity, respiration, growth, and the effects of turbidity on coral growth and metabolism. Kendall and Powell (1988) describe a STACH (short term aerated coral habitat) Plexiglas dome with inflow and outflow ports, which is capable of holding a number of coral branches. Scuba cylinders supply aeration, and radioactive tracers can be injected with a syringe through a stopcock. Carbon allocation in kelps has been evaluated using Plexiglas chambers (Buggeln and Lucken, 1979) as well as by applying ^{14}C embedded in petroleum jelly to corrugated blades and then surrounding the area with a sealed chamber (see Buggeln, 1985 for review).

Disease in corals is becoming more common and is a major factor in the declines in coral communities (Gochfeld and Aeby, 2008; Reed et al., 2010). Samples can be taken from corals using a syringe or scraper. A Diving PAM fluorometer can be used to assess the photosynthetic yield of coral's zooxanthellae. This is a non-destructive method that enables researchers to determine the health of the coral and its symbiotic partner under varying conditions *in situ*.

FIGURE 7.11:
Diving PAM fluorometer used to assess the health of coral.
(photo courtesy of D. Gochfeld)

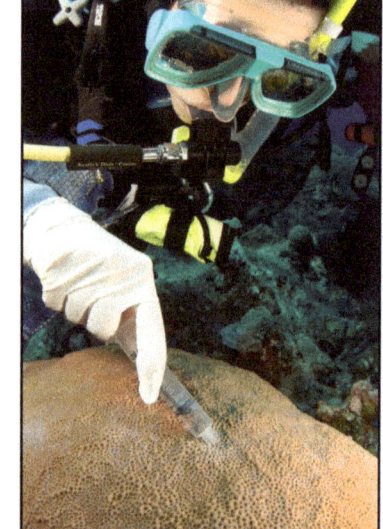

FIGURE 7.12:
Coral disease sampling using syringe .
(photo courtesy of A. Gelber, PBS&J)

FIGURE 7.13:
Measuring coral outplant *in situ* with ruler.
(photo courtesy of M. Miller)

Distribution and Abundance Studies

Benthic Organisms
Descriptive surveys of the distribution and abundance of subtidal organisms are commonly conducted by scientific divers. There is a large body of literature concerning these techniques, which should be consulted when developing a research plan (see Littler and Littler, 1985). Preliminary site visual evaluations and a pilot study may be necessary before a firm research plan is formulated. Consideration should be given to how many replicate samples are needed, assurance of proper randomization, and how the data will be analyzed, before significant effort is exerted underwater (see Andrew and Mapstone, 1987, for sampling review).

Shepard (1985) describes a method of free-range searching to quantify abalone over a wide area. He measured the "power" of a diver, which is defined as the effective area covered by the diver per unit time, and the "efficiency," which is the ability of the diver to discriminate between habitat types, and to find abalone in the correct type of habitat. He incorporates indices of surge and algal density, which affect a diver's ability to detect abalone while searching. This method is most useful for species that are distributed randomly, rather than aggregated in specific types of habitat, and for organisms that have relatively low densities.

Transect tapes and quadrats are often used to aid in the measurement of the distribution and abundance of organisms of limited mobility, such as algae and benthic invertebrates. Transect tapes made of flexible fiberglass material are available in various lengths [e.g. 15 m (50 ft), 30 m (100 ft), 50 m (165 ft), and 100 m (330 ft)] and graduations (e.g., cm/m one side and in/ft other side; see Forestry Suppliers ref.). Transects can be laid across some type of environmental gradient, such as depth, temperature, light, dissolved oxygen, or nutrient concentration, thus producing stratified samples. To avoid bias, transects should be laid out using random compass bearings.

Permanent transects can also be established in order to resample the exact same areas over time. These are often used for long-term ecological monitoring projects (Gittings et al., 1990). Leaded line, available from marine supply stores, can be laid along the bottom and attached at various points to mark a permanent transect (see below).

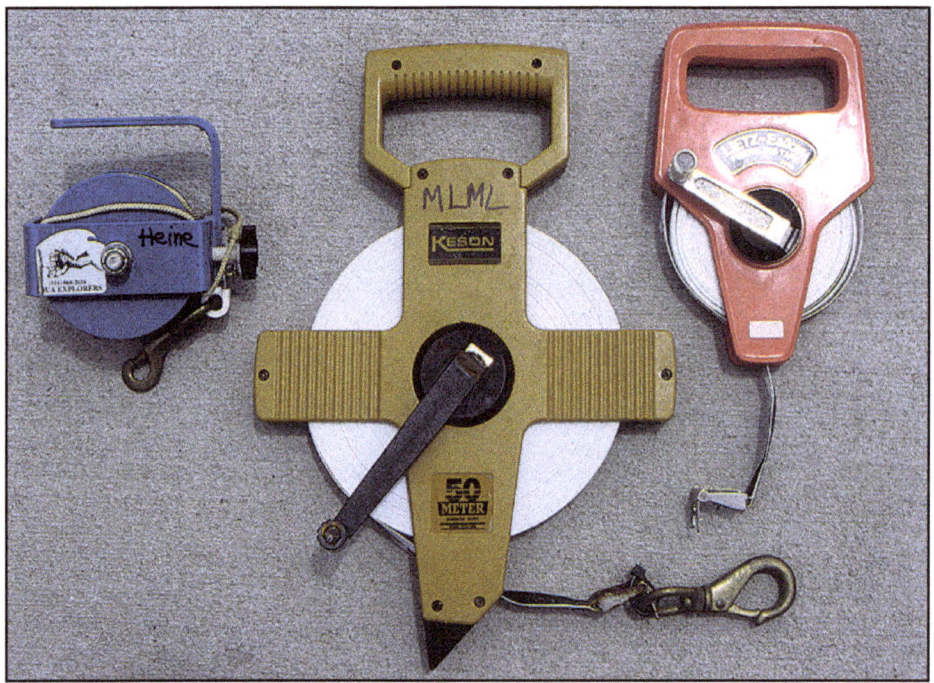

FIGURE 7.14:
Transect tapes and reels come in various lengths and styles.
(photo courtesy of J. Heine)

Sessile and mobile benthic organisms can be counted in a band or swath transect of pre-determined length and width, or by using discreet quadrats. A band transect often uses a length of transect tape (e.g., 25 m), where a diver swims a swath, perhaps using a 1 m stick held at a right angle to the transect tape, to delineate an area to be sampled (25 m x 1 m = 25 m^2). The other side of the transect tape can be sampled as well, yielding an additional 25 m^2 of area (Ambrose et al., 1993; Lessios, 1996; PISCO, www.piscoweb.org). Individual organisms can be counted and recorded on a slate. If large numbers of one species are to be counted, a hand-held tally counter is useful for increased accuracy and for limiting double counting or missing individuals.

Another variation is called the uniform point contact (UPC) method, which is used to estimate the percent cover of species as well as record the substrate type (i.e., sand, cobble, boulder, or bedrock) and physical relief (www.piscoweb.org). Along a 30 m transect, 30 points will be sampled.

Quadrats can be constructed from PVC pipe, aluminum, steel bars or plate, or other suitable material. The size of the quadrat must be commensurate with the size of the organisms being studied (see Chapman, 1985; Andrew and Mapstone, 1987). Quadrats are usually square, with common sizes ranging from .01 to 5 m^2. Variations on quadrat design include sub-dividing with monofilament or other fine line or wire to aid in visual estimation of percent cover, leaving out one side of the quadrat in order to place around large kelps or other organisms, and making corners hinged so that large quadrats can be collapsed for easier mobility. Quadrats must be heavy enough to stay on the bottom, and can be weighted with small pieces of rebar if necessary. PVC pipe quadrats can be drilled to allow water inside. Also, some researchers prefer to not glue the pipe and fittings, so the quadrat can be transported collapsed. Another variation runs elastic "bungee cord" inside the PVC pipe so that parts are not lost. Permanent quadrats can be marked at the corners with epoxy, nails, or bolts inserted into drilled holes.

Percent cover or counts of individuals can be made for the entire quadrat, or for each grid in a divided quadrat, or for the species beneath the intersections of the dividing lines

FIGURE 7.15:
A diver estimates percent cover using a divided quadrat on a reef wall.
(photo courtesy of M. Ponti)

within the quadrat (Doty, 1969). Visual estimations of percent cover can be made for specific functional groups or species, such as brown, green, or red algae, sponges, corals, bare rock, sediment, etc. Some researchers use percent categories such as 0, <1, 1-5, 6-25, 26-50, 51-75, and >75% (Chiappone et al., 1996).

Point-intercept methods can also be used to get a more precise estimation of percent cover, however, they usually require more time to collect the data. They can also be used to quantify layering of organisms, which can produce a percent cover of greater than 100% (Littler and Littler, 1985). One method uses a grid of lines within a quadrat, and each species or category of organism or substrate type that occurs directly underneath the point-intercepts is recorded on a slate or tape recorder (Chiappone et al., 1996). Another method, called random point contact (RPC), uses a weighted bar with a string of knots tied to each end. The bar is dropped into a quadrat, and each species or group under the string, and above it to a specified height, is recorded on a data sheet (Foster et al., 1985).

The PISCO group (www.piscoweb.org) uses three-sided 1 m² quadrats made of PVC. They are laid along the transect line using the transect as the fourth side of the quadrat. The diver records the number of all targeted species in the quadrat.

A line-point intercept transect method utilizes a weighted line with knots regularly tied along its length. Divers record the substrate, species, or type of organism found directly under each knot on the line (Liddell and Ohlhorst, 1987; Trunnell and Nelson, 1989). A variation on this method uses chain with links of approximately 1 cm, and the number of links with which each organism comes into contact is determined (Lessios, 1996). One

FIGURE 7.16:
A diver conducting random point contact (RPC) estimation of percent cover.
(photo J. Bodkin; courtesy of U.S. Fish and Wildlife Service)

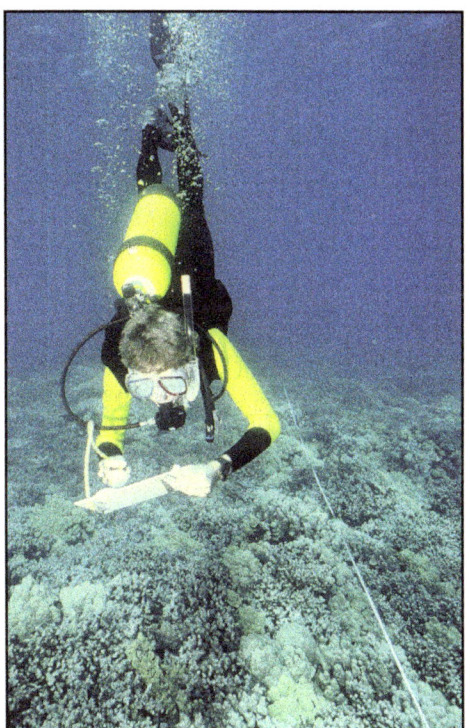

FIGURE 7.17:
The line-intercept method.
(photo courtesy of J. Pye)

SWATH / PISCO subtidal sampling / So Cal / rev. 06/06

SITE ________ Date ________ Observer ________ Buddy ________
Side: N S E W CEN Zone: I M O Transect: 1 2 Heading: Depth:

SPECIES	0-10 m	10-20 m	20-30 m
MACPYR(stipes)			

MACPYRAD	Macrocystis adult (>1m)	ASTMIN	Asterina miniata	MIMFOL	Mimulus foliatus
PTECALAD	Pterygophora adult (>30cm)	DERIMB	Dermasterias imbricata	CANSPP	Cancer spp.
CYSOSMAD	Cystoseira (diameter>6cm)	ORTKOE	Orthasterias koehleri	PUGPRO	Pugettia producta
LAMSET	Laminaria setchelli adult (stipe >30cm	ASTSPP	Unidentified Stars	PUGRIC	Pugettia richii
LAMFAR	Laminaria farlowii ad (width>10c	PISBRE	Pisaster brevispinus	LOXGRA	Loxorhynchus grandis
EISARBAD	Eisenia ad (>30cm)	PISGIG	Pisaster giganteus	LOXSCY	L. crispatus/S.acutifrons
ALAMAR	Alaria marginata	PISOCH	Pisaster ochraceous	PANINT	Panulirus interruptus
NERLUE	Nereocystis leutkeana	PYCHEL	Pycnopodia helianthoides	LITUND	Lithopoma undosum
COSCOS	Costaria costata	MEDAEQ	Mediaster aequalis	LITGIB	Lithopoma gibberosum
AGAFIM	Agarum fimbriatum	PARCAL	Parastichopus californicus	CRAGIG	Crassedoma giganteum
STRPURAD	S. purp(>2.5cm)	PARPAR	Parastichopus parvimensis	KELKEL	Kelletia kelletii
STRFRAAD	S. francisc(>2.5cm)	ANTXAN	Anthopleura xanthogrammica	CYPSPA	Cypraea spadicea
CENCOR	Centrostephanus coronatus	ANTSOL	A. sola	CRYSTE	Cryptochiton stelleri
LYTANA	Lytechinus anamesus	URTLOF	Urticina lofotensis	MEGCRE	Megathura crenulata
TETAUR	Tethya aurantia	URTSPP	Other Urticina	APLCAL	Aplysia californica
STYMON	Styela montereyensis	METSPP	Metridium spp	BALNUB	Balanus nubilis
MURCAL	Muricea californica	PACFIM	Pachycerianthus fimbriatus	HALIOTIS SPP	MUST ID SPECIES
MURFRU	Muricea fruticosa	LOPCHI	Lophogorgia chiensis		

Entered on ________ by ________ Checked on ________ by ________

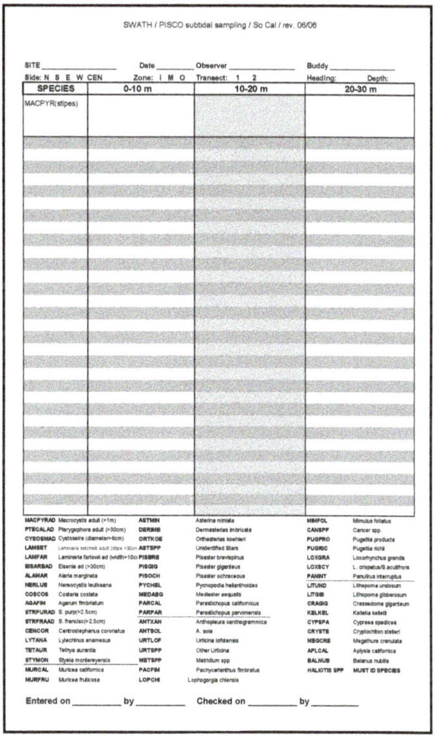

FIGURE 7.18:
An example of a benthic swath data sheet showing species codes of common biota that are counted .
(courtesy of UCSB/PISCO)

of the advantages of this method is that it accounts for the three-dimensional structure of the substrate and sessile organisms. Quadrats are two-dimensional, and percent cover or density of organisms may be over-estimated if the substrate is highly irregular (Dahl, 1973; Carleton and Sammarco, 1987). Lead line or chain can be used as a measure of substrate heterogeneity and rugosity (i.e. flatness vs. steepness), by using the ratio of measured chain length to the linear distance on a transect tape (Luckhurst and Luckhurst, 1978; Jaap et al., 1990) (see also Chapter 6).

Survey methods have been combined in a Baseline Survey Protocol (BSP) in order to conduct quantitative coral reef assessments (Markham and Browne, 2007). This technique uses four divers on a 45 m transect, which is split into two 2 m x 20 m sections with a 5 m gap. The divers have different roles, such as physical surveyor (water temperature, horizontal visibility, depth, current strength, water sampling, and rugosity), fish surveyor (species and size of all fishes seen, fish counts), benthic surveyor (line intersect transect method, substrate type), and invertebrate and algae surveyor (records all species of algae, takes samples, record abundance of invertebrates, photography, sediment samples).

Photo quadrats are often taken, with subsequent analysis in the laboratory. Dupont and Coy (2008) used digital photographs acquired every 0.5 m along a 15 m transect line, at a consistent distance of 0.5 m from the bottom. Twenty random points were scored using Coral Point Count v. 3.4 (Kohler and Gill, 2006) (see Chapter 8).

Jewett et al. (2008) used a nested quadrat design for benthic sampling in the Aleutian Islands. They took a photographic record (Canon PowerShot S80 digital camera) of a 1 m² quadrat, and inside of that was a 0.25 m² quadrat which was also photographed before all macroalgae were removed, and another 0.0625 m² quadrat was scraped free of flora and fauna.

Destructive sampling is done by collecting organisms within a quadrat. Chenelot et al., (2008) used scrapers to collect foliose and crustose algae in a 25 cm x 25 cm (0.06 m²) quadrat. Invertebrates were collected using an airlift suction dredge with a fine-mesh (500 μm) drawstring bag, powered by a scuba cylinder.

TRAINING EXERCISE #8
Sample Survey of Common Benthic Biota
Familiarization with Transect Tapes, Quadrats, and Slates

Equipment needed: transect tapes, quadrats, slates, compasses
Objective: To perform a simulated survey to be determined by the staff

The class can be divided into buddy pairs or groups of three. A thorough briefing must be given by the teaching staff prior to the dive. A "dry run" on land is often beneficial prior to the dive. Each group will have one compass, one transect tape, one quadrat, and one slate to record data on. Each group will be given a different random compass bearing so that they will not be in each other's way underwater, and random distances to sample along the transect. The dive groups will descend, and attach the transect tape to a fixed buoy line, the boat anchor, a kelp holdfast, or similar structure. They then swim the tape out on their compass bearing to the designated distance, which is the greatest random number assigned to be sampled along the transect. They will place the quadrat in a pre-determined location along the tape (e.g., the upper right-hand corner of the quadrat on the random number of the tape), and conduct a simple survey to be decided by the staff. Suggested topics include counting common large invertebrates or plants for density determinations, estimating percent cover of common invertebrates or algae, counting clam siphons, or whatever the local conditions allow. The following data can also be recorded for each quadrat: depth, substratum type (i.e., solid rock, cobble, gravel, sand, mud), and aspect/slope. As the divers proceed along the transect, they should rotate through tasks such as rolling in or out the transect tape, using the quadrat, and recording data, so that everyone has a chance to handle all of the equipment in the exercise. Each group of divers can be assigned different organisms to count, and the data can be pooled in class for evaluation and write-up. A debriefing is helpful to discuss problems encountered and to solicit suggestions for improvement or modification of the method for particular circumstances. An important part of this exercise is allowing the divers to plan and discuss together what they will do on the dive, and then evaluate it together with the staff after the dive.

Pelagic Organisms

For pelagic species of fishes or invertebrates, timed swimming transects can be used, or alternatively, a method of counting a number of fin kicks that have been calibrated to a known distance on a transect tape. One major source of error in these three-dimensional transects is identifying and adhering to a discreet volume of water to be surveyed (Biggs et al., 1981; Biggs et al., 1984). Underwater reference grids have been constructed from vertical lines suspended from a small boat, to which lateral lines are strung, to delineate an underwater volume (Biggs et al., 1986). By measuring the current, or by motoring a small boat at a slow pre-determined speed for a certain amount of time, a volume flow of water can be calculated and used for abundance estimates of organisms.

FIGURE 7.19:
Scientific blue-water divers preparing for a transect to estimate pelagic organism abundance.
(photo courtesy of N. Wu)

Aggregate volume of small pelagic organisms can be determined by macro-photography with a scale mounted on the framer (Trent et al., 1978; Hebel, 1983). Volume samples can be estimated *in situ* with a hand-held flow meter (Trent et al., 1978).

Video transects have been used with varying success, depending upon the resolution of the film or camera and the water visibility, and the ability of the investigators to identify organisms on video. See Haddock and Heine (2005) for complete description of equipment and techniques for blue-water diving (also see Chapter 3).

Fish

Quantifying fish populations using underwater transects has a wide and varied history in the literature (Brock, 1954; Sale and Douglas, 1981; Brock, 1982; Sale and Sharp, 1983; Lincoln-Smith, 1988 and 1989; Davis and Anderson, 1989; Deehr et al., 2007). The method must be designed with respect to the type of fishes being studied, their habitat, environmental conditions (water visibility, surge, time of day, etc.), and the effects of the presence of divers. There are two major categories of visual census techniques: those where the observer moves along a transect and counts fish species and numbers, and those where the observer remains stationary and records species and numbers of fish in a specified area for a specified time period (Lessios, 1996).

Many species of fish are not easily documented using traditional transect techniques. Various methods have been developed to adequately census fish populations, including a visual census technique (Russell et al., 1978), timed transects (Jones and Thompson, 1978), and stationary visual survey methods (Bohnsack and Bannerot, 1986; Clavijo et al., 1989; Dupont and Coy, 2008). For benthic species, band transects of pre-determined length and width can be used. For example, a buddy pair of divers can each swim a band transect along either side of a shared tape in between them, and still maintain acceptable buddy contact. Each diver records the species and number of fish encountered in a volume of water such as 2 m wide by 2 m above the bottom (Clavijo, 1990). Larson and DeMartini (1984) used band transects of 75 m long by 3 m wide by 1.5 m high, for a total "transect volume" of 337.5 m^3. Data can be recorded on a slate, or recorded audibly on a tape. The rate of swimming the transect should be standardized to account for fast swimming and cryptic species. Potential problems include fishes that move away from divers as well as

FIGURE 7.20:
A diver conducts a visual census of fishes on a rocky reef.
(photo courtesy of S. Hamilton)

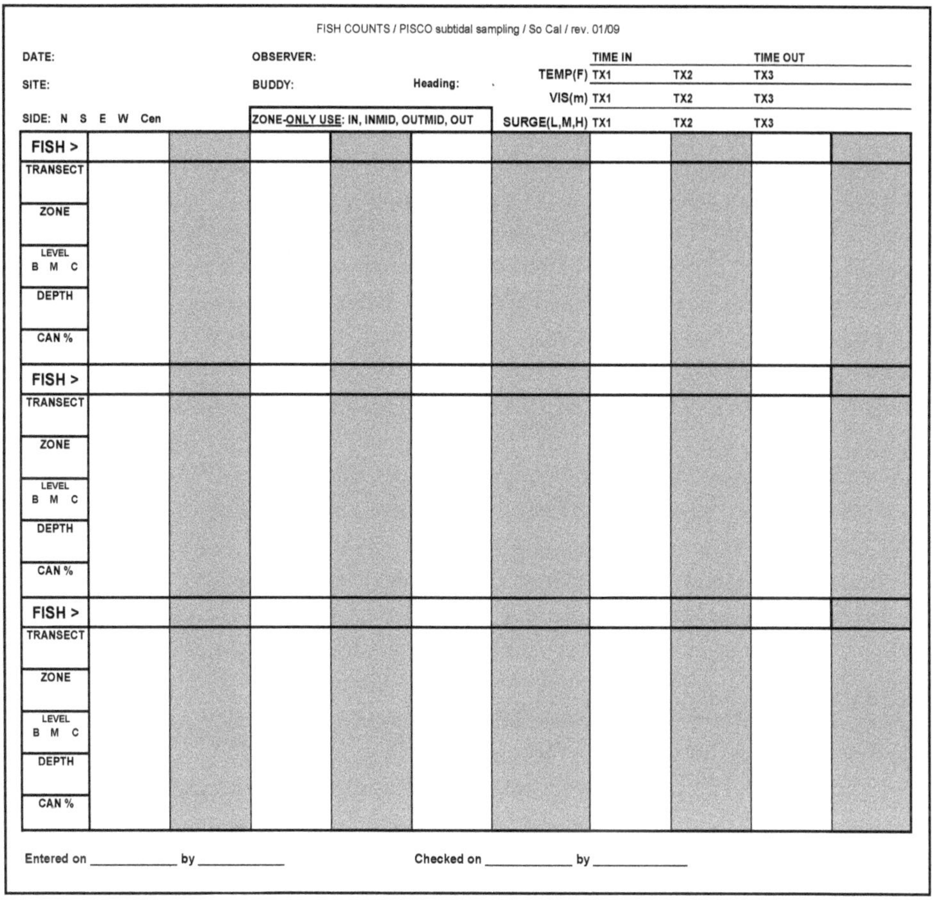
FISH COUNTS / PISCO subtidal sampling / So Cal / rev. 01/09

DATE: OBSERVER: TIME IN TIME OUT

SITE: BUDDY: Heading: TEMP(F) TX1 TX2 TX3

VIS(m) TX1 TX2 TX3

SIDE: N S E W Cen ZONE-ONLY USE: IN, INMID, OUTMID, OUT SURGE(L,M,H) TX1 TX2 TX3

FISH >										
TRANSECT										
ZONE										
LEVEL B M C										
DEPTH										
CAN %										
FISH >										
TRANSECT										
ZONE										
LEVEL B M C										
DEPTH										
CAN %										
FISH >										
TRANSECT										
ZONE										
LEVEL B M C										
DEPTH										
CAN %										

Entered on ____________ by ____________ Checked on ____________ by ____________

FIGURE 7.21:
PISCO subtidal fish counts data sheet.
(courtesy of PISCO)

fishes that might be attracted to divers, biases due to water clarity or surge, biases due to divers' estimation of defined area to be surveyed, repeat counting of the same fish, and diver experience in fish identification.

Reef Check California uses 30 m x 2 m x 2 m transects to count the density, size distribution and sex (where possible) of reef fishes (www.reefcheck.org). The PISCO group (Partnership for Interdisciplinary Studies of Coastal Oceans) has a long-term monitoring program along the west coast of the U.S. (www.piscoweb.org). They use visual surveys to quantify the species composition and size structure of fishes in the kelp forest, and stratify the sampling to include bottom, mid-water, and canopy, as well as inshore and offshore. A transect of dimensions 30 m x 2 m x 2 m is sampled by a diving buddy pair. Four divers can complete 24 transects at all three depths at a site in one day. Visual strip transects of 25 m x 4 m x 4 m were used in Washington for groundfish abundance and size estimations (Weispfenning et al., 2005; Valz and Dinnel, 2007). One diver would count and measure groundfish, using a 20 cm wide dual-laser (Lasermate, Inc.) measuring device, to the nearest 5 cm. The second diver followed the first diver and estimated bottom composition (sand and gravel, rock, boulder, and wall). Transect width is an important factor when designing a fish survey methodology (Cheal and Thompson, 1997; Sale, 1997).

The visual census technique (Russell et al., 1978; Bohnsack and Bannerot, 1982) is non-disruptive, repeatable, and adaptable to a variety of environments and conditions. However, with many researchers modifying the sampling design with respect to area surveyed, sampling time, and other protocols, this has made comparisons between studies difficult (Bortone et al., 1992), and many biases may be present (Sale and Douglas, 1981; DeMartini and Roberts, 1982; Sale and Sharp, 1983; Watson et al., 2010). This method requires that divers occupy the center of a circle, for a predetermined period of time, usually 5 minutes, count the number of individual fish by species (Bohnsack and Bannerot, 1986; Clavijo et al., 1989; Dupont and Coy, 2008). Sizes of the circle that have been used are 5.6 m (20 ft), or 7.5 m (25 ft) radius, The size of the visual area can be adjusted depending upon water conditions. The sampling area can be stratified by depth zones, with divers conducting visual surveys working from deep to shallow (Scarborough-Bull and Kendall, 1992). One method uses timed transects, where divers record the species and number of fish observed during a discreet time period (Jones and Thompson, 1978).

The Roving Diver Technique (Schmitt and Sullivan, 1996) is a non-stationary survey method where free-swimming divers randomly roam through a designated study area, recording all fish species seen. After the dive, other information is noted, such as feeding behaviors, assignment to specific abundance categories, water visibility, and other abiotic factors (Postell et al., 2008).

A combination of census methods can be employed to overcome biases with one particular method and to record non-mobile, cryptic species. Both transect surveys (30 m x 2 m) and radial point counts (7.5 m radius circle) were used in Belize to establish an Ecopath network (Deehr et al., 2007).

A visual fast count estimator of fish abundance has been applied to underwater video sampling (Barry and Coggan, 2010). The goal was to reduce the bias inherent in diver visual estimations. Baited remote underwater stereo-video and diver operated stereo-video transects have also been used effectively (see Langlois et al., 2010, for complete review; also see Chapter 8). Stereo-video techniques reduce inter-observer variability, increases accuracy of fish length estimates, and improves definition of the sample unit area.

Towed-diver surveys using a "manta tow board" are useful for surveying large areas rapidly (Boland and Donohue, 2003; Boland, 2005) (see also Chapter 3 for description of method).

Fish mortality occurs when oil platforms are removed from service by the use of explosives. Researchers from the National Marine Fisheries Service have developed sampling protocols to assess the impacts of these explosives on fish communities (Gitschlag, 1995). They used 3 m x 3 m frames with 5 cm trawl mesh netting that collected dead fish that sink to the bottom after the explosion. The frames were deployed and retrieved by divers, and transects were run to collect dead fish by hand. An open water purse net was also used, but required considerable diving time to deploy and retrieve. This method was replaced by the use of 6.7 m diameter circular surveys where divers collected dead fish, which were placed into mesh bags.

TRAINING EXERCISE #9
Fish Transects

Equipment needed: transect tapes, slates, data sheets, compasses, depth gauges
Objective: Orientation and practice using equipment and fish identification

The class can be divided into buddy pairs. A thorough briefing must be given by the teaching staff prior to the dive, which includes the dive plan and the use of each piece of equipment. Fish identification must be covered prior to the dive.

Suggested transect lengths are 25 x 2 x 2 m, which can be adjusted according to water visibility and conditions. Divers will swim on either side of the transect tape and record the number and size of fish species encountered. A visual census technique may also be used, where timed surveys of a certain dimension are conducted with stationary divers. The class may be able to assist a local researcher with their research project.

Collecting

THE COLLECTION OF ORGANISMS IS DONE for many scientific reasons, including studies of anatomy, physiology, behavior, ecology, and curation in museums. Many organisms cannot be collected by remote means (i.e. grabs, towed nets, trawls, etc.) due to their fragile nature, low density, or evasive behavior. Divers are often able to collect these species with many ingenious methods, which will be covered by habitat and general species type below.

Collecting permits are required in many areas of the world. These permits are often specific to organisms allowed to collect, and tools and methods permitted to use. It is best to check with local governmental agencies prior to collecting. Collecting of venomous or toxic organisms must obviously be done with extreme caution. A valuable reference to hazardous marine life is Auerbach (1997).

Pelagic Organisms
Collecting small organisms in blue water/open ocean conditions can be quite challenging (Haddock, 2004). Gelatinous zooplankton can be collected by hand with wide-mouth glass jars between 470 ml (16 oz) and 1 liter (32 oz) with screw-on lids (Mann, 1989; Haddock and Heine, 2005). All jars, bottles, and bags should be pre-filled with water before lowering underwater, and either have the lids loosely screwed on, or removed. Planktonic fora-

FIGURE 7.22:
A blue water diver collects pelagic organisms in a glass jar. (photo courtesy of S.H.D. Haddock)

minifera have been collected in shallow water (depth of 3-5 m/10-16 ft) with wide-mouthed glass jars (130 ml) or small plastic snap-cap vials, by sighting them against the dark underside of the diving vessel (Huber et al., 1996). The best method of collection is to gently rotate the jar around a specimen, while slowly lowering the top of the jar. Plastic bags can also be used, but they tend to be bulky and have considerable drag (Biggs et al. 1986).

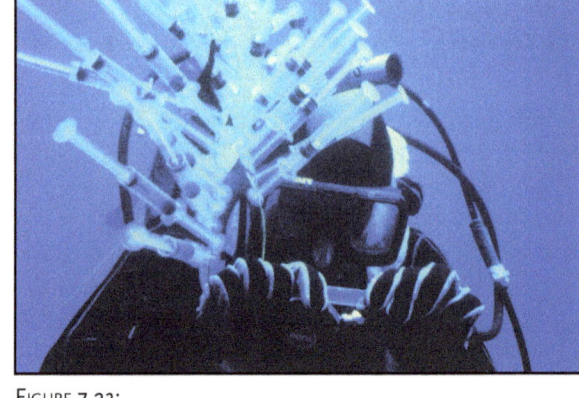

FIGURE 7.23:
A blue water diver collects marine snow particles with syringes.
(photo courtesy of K. Coale)

Samples of small organisms can be taken with a plastic or glass jar fitted with a two-hole rubber stopper. A flexible piece of tubing is fitted into one hole. The diver inverts the open jar and purges it with air, inserts the stopper, and when the jar is righted, air bubbles come out of the open hole and water enters the sampling tube. When the container is filled, the flexible hose end can be inserted into the open hole on the stopper to cap the container (Miller, 1975; Somers, 1990). Marine snow has been collected by using an automatic pipetting syringe with a long coil of Tygon tubing attached (Hebel, 1983; Alldredge and Silver, 1988). This allowed for discreet sampling and the collection of sufficient quantity for analysis. Divers can also push or tow small plankton nets or other types of nets through the water. Some researchers have also attached nets to diver propulsion vehicles (Ennis, 1972; Schroeder, 1974). Cod ends can be changed underwater, and the samples can be preserved by injecting formalin through a thin plastic film covering the jars. An overview of collecting and handling zooplankton and epibenthic organisms can be found in Schroeder (1974).

Bacteria

Samples of bacteria have been collected from corals *in situ* using a sterile 3 ml syringe (Richardson, 1992) (see Figure 7.12). Another method of taking bacterial samples underwater from larger invertebrates without handling the animal is described by Forsythe et al., (1993). They slowly approached

FIGURE 7.24:
A diver prepares to collect multiple samples of exuded chemicals from a soft coral using syringes.
(photo courtesy of J. Heine)

octopus, swabbed the surface of the animal, and inoculated an agar slant underwater. The culture media was transported underwater in screw top tubes filled with sterilized mineral oil. The tubes were held in a PVC tube-holder (4 in. PVC), which could carry up to 8 sample tubes. The tubes were held upside-down when the screw cap was removed, which kept the mineral oil from floating out of the tube, as well as preventing sea water from floating in.

Infauna

Infaunal organisms are often collected and quantified by taking core samples. Containers can range from various sized metal cans, to plastic or glass jars, to PVC pipe cylinders. Containers are often modified by cutting out one end, and affixing fine mesh netting (e.g. nytex) to allow water (but not the desired organisms) to escape as the core is being pushed into the sediment (with the netting end at the top). The top or lid of the container must be slid alongside the container and put into place before the core sample is removed from the substrate (Coyer and Witman, 1990; Coyer et al., 1999). Various types of custom-made trays (e.g. six packs) have been designed to hold core samples. Depending upon the size and number of core samples collected, a lift bag or line to the surface may be required to move the samples to the boat or to shore. Other types of corers have been constructed of PVC pipe (schedule 80), with a piece of wood with a handle on top. The corer is pushed into the

FIGURE 7.25:
Infaunal organisms can be easily collected using cores with end caps.
(photo courtesy of C. Yanch)

substrate, and a rubber stopper or cap is placed onto the open end of the corer before it is removed with the handle. A multi-level corer, which samples an area of 45 cm^2 to a depth of 6 cm, is described by Fager et al., (1966). It has a square brass box and slots for thin metal slide plates to be inserted into a sample. An Ekman grab (15 cm x 15 cm) can be pushed into the sediment by divers, and the samples can be transferred to plastic bags underwater, allowing for multiple samples to be taken during a dive (Cahoon et al., 1992) (see Chapter 6).

Larvae and Demersal Zooplankton
Invertebrate larval recruitment is often measured using passive larval collectors or larval tube-traps (Butman, 1986; Yund et al., 1991; Powers et al., 1995). Designs usually use high aspect ratio tubes (60 cm/24 in. height to 5 cm/2 in. diameter), often with a layer of formalin in the water, to prohibit resuspension or escape of larvae. Collection tubes can be deployed on the bottom, or throughout the water column on moored arrays (Powers et al., 1995). The tubes can be capped and replaced underwater by divers.

Demersal zooplankton can be collected in a variety of ways, including vertical plankton net tows, reentry trapping, emergence trapping, airlift sampling, and Ekman grab sam-

FIGURE 7.26:
Horizontal plankton sampler (HOPLASA) collects near-reef plankton.
(photo courtesy of K. Rutzler)

pling, but each method has significant biases toward different groups of animals (Cahoon et al., 1992). Tronzo et al. (1986) describe the use of reentry traps which are 21 cm by 21 cm plastic pans filled with clean sand, and emergence traps, which are 1 m high conical mesh nets with a catch bottle on top, which are deployed by divers and left out overnight. Emergence traps have also been used to capture zooplankton as they emerge from the bottom (Hobson and Chess, 1979; Rutzler et al., 1980; Youngbluth, 1982). Klos et al., (2005) used diver-controlled plankton nets while tethered to a blue-water diving rig.

FIGURE 7.27:
Diver placing algal samples into plastic bags.
(photo courtesy of J. Heine)

Algae

Very small organisms can be located with the use of an underwater magnifying device. Kennelly and Underwood (1984) report constructing an incident-light microscope that they used to determine the relative abundances of micro-organisms living on natural substrata. Mladenov and Powell (1986) describe a simple lens unit, with a swiveling mirror, and light source. Macroalgae are easily collected by hand, or using simple tools such as a knife, scraper, hammer, or chisel. For a complete review of handling and preserving algae, see Tsuda and Abbott (1985). Algae can be placed into mesh or cloth bags underwater. Plastic bags can be difficult to handle underwater, but may be preferable for delicate specimens. See also Littler and Littler (1985) for a detailed description of phycological methods.

Macroinvertebrates

Invertebrates can BE collected by hand or with small hand tools, and placed into plastic, burlap, or mesh bags. Small, delicate organisms can be placed into numbered plastic vials or Ziploc® bags. Various types of traps have been used to catch crabs, lobsters, and octopus, some of which are similar to those used for catching fishes (see following section). Octopuses have also been collected from their holes by introducing bleach from small syringes into their dens. Crayfish and other macroinvertebrates were collected with a diver-operated suction gun and collection bucket (Davies and Ramsey, 1989). They reported that using the suction gun was easier, faster, and less damaging to specimens than collecting by hand. The collection bucket is made from clear acrylic plastic, with pie-shaped thin cellulose acetate butyrate wedges that form a self-closing top.

Cores can be taken from live coral colonies by use of a pneumatic drill powered by a scuba cylinder (see drilling, Ch. 4). Richardson and Carlton (1993) used a one-inch hole saw drill bit to take cores from a *Montastraea annularis* colony. In order to prevent infection, the resulting hole in the coral colony was plugged with modeling clay.

Another method of collecting a quantitative sample of macroinvertebrates down to a size of approximately 5 mm is by use of an oyster sampling apparatus (Ducey, 2009). A double layer mesh lining made from charcoal fiber window screening was placed into a standard mesh laundry bag inside a cube box (29 cm x 29 cm x 35 cm). Excavated sediment and invertebrates were placed by hand into the lined crate.

Fish

Many methods of collecting fishes have

FIGURE 7.28:
Small cores of coral can be obtained with a pneumatic drill.
(photo courtesy of PBS&J)

FIGURE 7.29:
Collecting macroinvertebrates with a suction dredge (see also later section).
(photo courtesy of M. Ponti)

FIGURE 7.30:
Juvenile and small fishes can be collected using a hand net.
(photo courtesy of N. Schiel-Rolle)

FIGURE 7.31:
Captured fish can then be transferred to a plastic bag for transport to the surface.
(photo courtesy of D. Kesling)

been learned from commercial, sport, or subsistence types of fishing, and range from simple dip nets to complex traps and even explosives (see von Brandt, 1964; and Baldwin et al., 1996 for complete reviews). Collection of live fish can be accomplished in a variety of ways. Small fishes can be caught with a hand-held dip net and placed in plastic or mesh bags. Sometimes divers use two nets, one for chasing small fish into the other. Flashlight fishes have been captured by divers at night under blackout conditions by stunning them with a flashlight beam (Baldwin et al., 1996). Small schooling fish, or juvenile or larval fish can be captured by one or more divers swimming a plankton net, or larger mesh size net (with an optional codend, or collecting jar, at the end of the net) through the water. Another modification of this method employs a net on the front of a diver propulsion vehicle. Lift nets use vertical lifting by divers, or personnel in a boat, and can be baited or illuminated at night to attract fishes. Lift nets (9 m^2) can be filled with organisms collected by divers who can use lift bags for moving the nets around and up to the surface (Scarborough-Bull, 1993).

Hoop nets are funnel-shaped, and have

FIGURE 7.32:
Wall net used to capture reef fish by herding technique.
(photo courtesy of K. Clifton)

CHAPTER 7 MEASURING BIOTIC FACTORS AND PROCESSES **155**

netting that is stretched around a series of decreasing diameter hoops that trap fish in the small end. They can be placed in discreet locations by divers. A Fyke net uses long "wings", which resemble fences, that guide fishes toward the opening of the net (von Brandt, 1964; Clifton, 1996). BINCKE nets, which are rectangular nets on PVC frames with Tygon tubing hinges, have been used to capture young fish (Anderson and Carr, 1997). Nets should not be left unattended on reefs, as non-targeted species can be trapped and killed.

There are a number of different types of traps made from almost any type of material that can be used to catch fishes or invertebrates. Simple types taper to a small end, which is usually baited, where the targeted fish species become trapped. Many other types of traps use a funnel valve, which is large on the outside of the trap, and small on the inside, which makes fish unaware that they are entering, and also makes it difficult to find the exit point. Special tags are available to mark traps (Floy Co.). Divers can also bait jars or other devices, and watch from a distance as fish swim into the device, and then swim over and cap it. Some traps are remotely operated (Smith et al., 1981). Garden eels are particularly difficult to catch, and have been snared by divers using a loop around the burrow of the eel. By backing off and waiting for the eel to emerge, the diver can pull on the end of the noose line, capturing the eel (Baldwin et al., 1996).

Divers have used hook and line, usually on a hand line or small fishing pole to catch fish underwater (Wilson, 1981). A "slurp gun" is a small acrylic tube with a plunger that uses suction to trap a fish in the tube, where it can be transferred to a bag.

For larger fishes, or fish that are in inaccessible habitats, chemicals can be used that anesthetize or stun. One example is quinaldine (2-methyl quinoline), which is mixed with

FIGURE 7.33:
Diver squirts quinaldine into hole to capture damselfish with hand net.
(photo courtesy of N. Schiel-Rolle)

alcohol (10% quinaldine to 90% methanol, Christie et al., 2010) and placed into a squirt bottle, large syringe, or plastic bag that can be perforated underwater (see Moring, 1987). Divers can squirt the diluted mixture directly onto desired fishes, or into holes, cracks, or crevices, and then collect the stunned fishes by hand, or with a dip net. Other anesthetics include MS-222 (Tricaine; Wedemeyer, 1970), Benzocaine (McErlean and Kennedy, 1968), clove oil (Perdikaris et al., 2010), and Metomidate. These chemicals generally work very fast, and usually have a good success for recovery of the fishes (Gilderhus and Marking, 1987).

Rotenone is a natural insecticide poison that can be purchased from farm suppliers. Five percent powder can be mixed with water to form a "paste" or "soup" that is placed into sealed plastic bags. It can be released by punching a small hole in the bag and squeezing the liquid out into the collection area (Smith, 1973; Baldwin et al., 1996). It generally takes about 5 to 30 minutes to become effective, but this is highly variable and dependent upon such factors as currents, temperature, mixing, and the resistance of fish species. The rotenone will usually kill the fish, and may also kill other unwanted species. It is generally not toxic to humans, but care should be taken when mixing to not inhale the fumes or come into direct contact with the poison. It may also leave a residue for some period of time (Engstrom-Heg, 1987), and produce other undesirable effects (Jaap and Wheaton, 1975).

Electrofishing uses a strong current of electricity to shock fish, which can then be easily collected by hand. It has been used extensively in freshwater streams, but is rarely used by divers for safety reasons. In freshwater applications, the main unit remains in the boat or on shore, and a diver places electrodes close to the fish and activates the device (James et al., 1987; Coyer and Witman, 1990). For marine applications, the topside unit has long

FIGURE 7.34:
Venemous fish such as this lionfish must be handled carefully.
(photo courtesy of M. Hay)

FIGURE 7.35:
Tagged damselfish beside PVC pipe used to collect fish eggs. Cow ear tag attached to rebar is used to identify the site.
(photo courtesy of N. Schiel-Rolle)

strands of copper welding wire that connect to the electrodes. When a desired fish swims into the field, the diver informs the crew on the surface to activate the unit (Davis and Anderson, 1989).

To collect fish that do not need to be kept alive, a simple pole spear, also known as a Hawaiian sling, is an inexpensive and relatively easy way to spear small to medium sized fish. Very small fish can be speared with custom-made pole spears using a stainless steel shaft and fine spring-steel wire embedded in epoxy cement at the tip. Larger specimens may require the use of a speargun, which come in a variety of sizes and power configurations, as well as the choice of different speartips. These spearguns are powered by heavy-duty rubber sling bands, or pneumatically with air (Barsky, 1997).

FIGURE 7.36:
A Wilson trap that attaches to a diver propulsion vehicle, and rebreather, used to capture sea otters.
(photo courtesy of M. Harris)

To collect damselfish eggs, Christie et al., (2010) placed a tagged damselfish beside a piece of PVC pipe (2" diameter) that is used by the fish to lay eggs. They placed a sheet of acetate paper (like that used for overhead projectors) inside the pipe for the fish to lay eggs on. The acetate can be pulled out, and a separate (clean) sheet of acetate paper laid on top and then traced around the egg cluster. If the sizes of eggs are known by the study species, then estimates of the number of eggs and fecundity can be made.

Mammals
Scientific divers have captured unsuspecting sea otters by using rebreathers and a specialized trap attached to a diver propulsion vehicle (Sanders and Wendell, 1991). They are able to sneak up on otters that are resting in the kelp canopy from below by using rebreathers which do not emit any bubbles (Canestro, 1997). Most collections of marine mammals are not done by divers.

Air Lifts, Suction Dredges, and Suction Samplers
Air lifts, suction dredges, and suction samplers are used to collect benthic invertebrates and algae. A number of designs for airlift or suction sampling devices have been reported by researchers. Most rely on the principle that air expands in volume as it ascends due to a decrease in surrounding pressure (Boyle's Law). The air used for the suction is usually provided from a standard scuba cylinder, or in some cases from the surface. One design houses the airlift in an acrylic box which has a flexible skirt on the bottom to conform to the substrate. This design prevents any organisms from escaping, thus assuring a valid quantitative sample (see Runnels, 1985 for details). These devices essentially vacuum up organisms from the substrate.

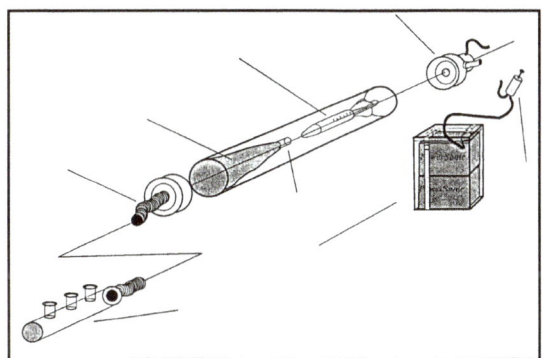

FIGURE 7.37:
Portable, diver-operated plankton sampler.
(from Sebens et al., 1992 AAUS Proc.)

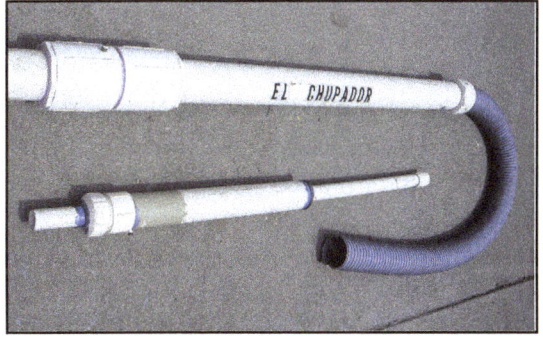

FIGURE 7.38:
Suction dredges can be made in various lengths and diameters.
(photo courtesy of J. Heine)

Another design for sampling demersal zooplankton is described by Sebens et al., (1992). This sampler is built from schedule 40 PVC pipe that is 80 cm long and 10 cm in diameter. At one end is a 12-volt submersible bilge pump, which is affixed to the pipe. A digital flow meter (General Oceanics) is mounted on threaded rod in the middle of the pipe. A piece of the pipe is drilled out, and a piece of Plexiglas is cemented in, to allow

FIGURE 7.39:
Collecting invertebrates on rock wall with underwater suction airlift powered by scuba cylinders, with nylon collection bag at the top.
(photo courtesy of C. Bianchi)

for flow meter readings. The other end of the pipe is fitted with a removable reducing coupling, with a small plankton net and collecting bottle inserted inside the pipe. The coupling has a flexible plastic hose that is used to collect the samples. The sampler is powered by 12 volt batteries in a watertight housing.

Airlift samplers are used to collect small invertebrates directly from the substrate. One such device is a 5 cm diameter PVC pipe that is 1.3 m in length, with an air inlet 40 cm from the bottom end (Cahoon et al., 1992). A scuba cylinder with first stage regulator and intermediate pressure is used to introduce air into the pipe, which creates suction as the air rises and expands in volume. A 105 μ mesh collection bag is affixed to the top of the pipe to collect sediments and animals. For a quantitative sample, a quadrat should be placed on the bottom to delineate the area to be sampled. Stretch (1985) compared quantitative sampling of demersal zooplankton with re-entry and airlift dredge techniques, and found marked differences between techniques.

A quantitative sample of benthic invertebrates was collected using a gasoline-powered irrigation pump, which provided 145 gal/min through hoses to a suction head (see Brook 1979 for description of airlift; Deehr et al., 2007). Collection bags had a mesh size of 2 mm. Samples were collected using a bottomless 20 L bucket with 30 cm inside diameter as a "quadrat."

A large volume surface-to-bottom airlift system is described by Pollock and Bowser (1995). The pipe was constructed of 7.6 cm (3 in) diameter PVC in ca. 5 m (16 ft) sections, with flexible swimming pool hose attached to the lower end for collecting. The system is powered by a set of twin 72 ft³ scuba cylinders, with a first stage regulator and low pressure hose. This system was used to collect foraminifera at depths of 30 m (100 ft). An anchor weight (ca. 11

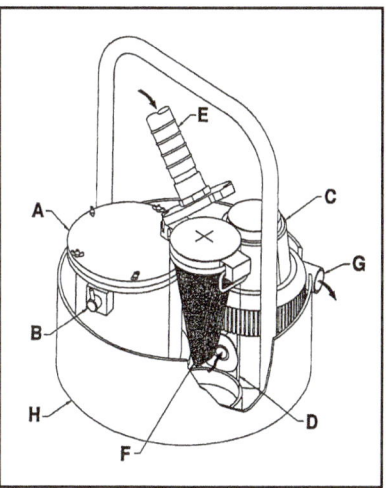

FIGURE 7.40: Schematic design of portable suction sampler "Priscilla." (from Miles and Whitlatch, 1997 AAUS Proc.)

FIGURE 7.41:
Portable suction sampler "Priscilla" being used in seagrass bed.
(photo courtesy of M. Dardeau)

to 23 kg/ 25 to 50 lb) was used on the bottom of the pipe to counteract the buoyancy of the volume of air rising through the pipe.

A suction sampler for collecting small subtidal gastropods (2-14 mm) from algal covered rocks is described by Miles and Whitlatch (1997). They use a modification of the design described by Taylor et al., (1995). A battery-powered water pump (West Marine 2000 GPH 12v bilge pump) allows for variable speed capabilities, and mesh collection bags can be changed underwater for continuous operation. See Figure 7.40 and Miles and Whitlatch (1997) for complete specifications and details.

Tagging Organisms

WITH THE WIDE VARIETY of types of organisms in nature, there are almost an unlimited number of methods of tagging that have been used, or could conceivably be invented. Some of the more common and useful methods are described here. It is important to test the effects of tagging, such as mortality, tag loss, and behavior changes in the laboratory before embarking on a large scale tagging effort.

Algae

Plastic cable ties (tie wraps) are available in a wide variety of sizes and colors, and can be used around the base of algal thalli, or in the holdfast of larger species. Small tags, such as pieces of plastic that have been etched with a soldering iron, or written on with permanent ink, or Dymo labels, can be attached to the cable tie before it is closed. Other useful tagging materials include surveyor flagging tape or bicycle handlebar tape, which can be tied around the plant and written on with permanent markers. These types of tags have been reported to persist in Nova Scotia kelp beds for more than two years (Chapman, 1985). Small thalli have been marked with bird bands (Barilotti and Silverthorne, 1972). Some researchers also tag the substrate with an appropriate marker so that they will know if the plant has been torn out or removed by grazers. Kelps have been marked with ink tattoos (Nicholson, 1970), or unique patterns or combinations of holes punched in the thalli.

FIGURE 7.42:
Diver measuring *Halimeda* tagged with a cable tie.
(photo courtesy of H. Spalding)

Invertebrates

As there is a wide diversity in invertebrate shape and form, there is also a wide variety of tagging techniques. For encrusting, non-mobile forms, it is often best to mark the substrate right next to the organism. Sponges and other benthic, relatively flat forms may be marked for identification by pounding a numbered nail or disk on a nail into the substrate next to the organism. Areas of coral have been tagged using nails inserted into the corals. Small numbered floats can be attached to the nails for specific identification. Coral growth, or necrosis due to disease, can be measured using a ruler or close-up camera between nail markers (Richardson, 1992).

Mollusks with hard shells can be tagged in a number of ways. Small bivalves (0.8 to 1.2 cm length) have been marked for recapture by affixing pre-printed and color-coded bee tags (2 mm diameter; see Bee Works at end of chapter) to the shell with Z-Spar marine epoxy (Zardus, 1997). Floy Tag, Inc., also makes vinyl tags that adhere to shells. However,

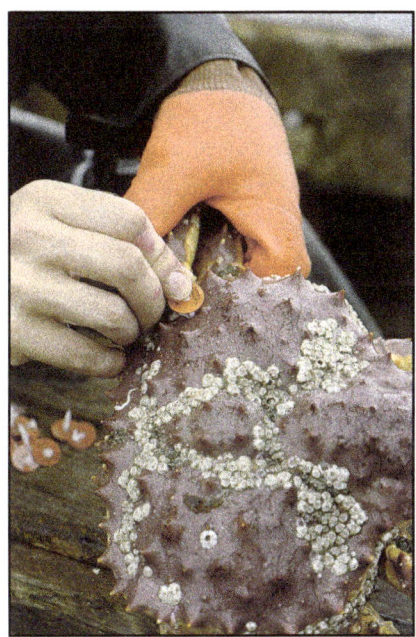

FIGURE 7.43:
Applying a tag to the carapace of a crab. (photo courtesy of W. High)

this is very time consuming. Scallops have been successfully marked using quick-setting calcium carbonate cement (Hudson, 1972). The shells of mollusks can also be written on with permanent ink, india ink, or fingernail polish (Hidu and Hanks, 1968). It may be necessary to lightly sand the shell before applying the marker, then cover with resin or clear cement for protection (Coyer and Witman, 1990). Shells may also be able to be engraved with a small grinding tool. Gastropods were tethered by gluing monofilament line with epoxy to the shells, in an experiment to see if they had a refuge amongst corals (Herrlinger, 1983). Mollusks with holes in their shells, such as abalone, can be tagged with stainless wire and numbered discs. In order to monitor the nocturnal movements of abalone, a small battery powered light was attached to its shell (Tutschulte, 1968). Other organisms that reside in soft-bottom sediments can have markers such as PVC pipe, rebar stakes, or plastic spoons placed adjacent to their siphons or burrows.

Lobsters have been tagged with color-coded tags inserted into the dorsal musculature between the abdomen and thorax with a syringe needle (Cooper, 1970). If a small hole (ca. 4 mm) is punched into one of the tail fan sections, the mark may be retained through at least one molt cycle. Crabs have been tagged with spaghetti tags between the carapace and abdomen, or around the legs. Blue crabs have been tagged by using a tag on a stainless steel wire leader that is attached around the spines on the carapace. Floy Tag, Inc., has crustacean tags specifically designed for shrimp, lobster, or crayfish (see Floy ref.).

FIGURE 7.44:
Sea urchins tagged with surgical tubing on their spines, and their home scar tagged with marine epoxy and numbers. (photo courtesy of L. Rogers-Bennett)

Soft-bodied invertebrates can often be tagged with Floy tags inserted with a tagging gun. Kline and Scheel (1996) report tagging octopuses with spaghetti tags, as well as sonic tags which were used to track the individuals. They also report using a diver-held receiver for *in situ* relocation of tagged octopuses, but do not give any details on the instrument. Octopus arms have also been tagged with a pin that has discs on each end, as well as with branding and staining techniques (Kayes, 1974). High (1998) reports tagging octopus with two dart tags and one loop tag on the animal's body. Sea anemones have been marked on their column with red dye applied with a paint brush (Sebens, 1976). Polychaetes have been tagged by imbedding microwire inside the animal and recapturing individuals by using a metal detector (Joule, 1983).

Sea urchins have been tagged in a variety of ways (Ebert, 1965; Lees, 1968). One method uses stainless steel wire and color-coded beads that are inserted through the test (Nelson and Vance, 1979). Floy tags have also been glued to urchin tests with silicone (Olsson and Newton, 1979), and inserted through a carefully drilled small hole in the test (Neill, 1987). Elastic bands with tags have been stretched over urchins (Dance, 1987). Plastic straws (Mattison et al., 1977), thin rubber or latex discs (Carpenter, 1984), and surgical tubing have been placed onto urchin spines. Lastly, wire harnesses have been bent into a pentagonal shape with long "tails" and loops, and placed around the urchin, with the tails bent up through loops in the wire to secure the harness (Coyer and Witman, 1990).

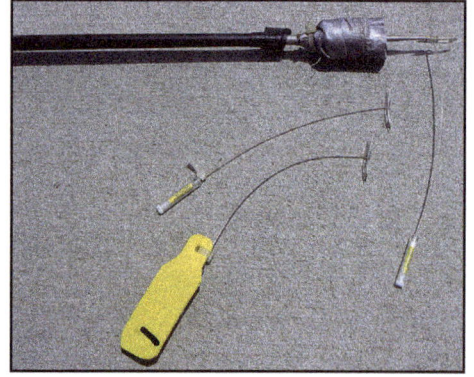

FIGURE 7.45A:
A pole spear can be used to place tags into larger fishes.
(photo courtesy of J. Heine)

Similar types of wire harnesses have been used for the harder, tougher species of seastars as well. Stains such as Nile blue sulfate have also been successfully used (Feder, 1955; Campbell, 1984). Floy and spaghetti tags inserted through the ray are particularly effective and long-lasting in many species of sea stars (Birkeland, 1974; Dayton et al., 1974; Savy, 1987).

Fish

Choice of tagging methods for fishes is dependent upon many factors, such as the size, color, and skin type of the fish, and visibility in the field (Jakobsson, 1970; Clifton,

FIGURE 7.45B:
Rockfish tagged with Floy plastic anchor tag.
(photo courtesy of S. Halewood)

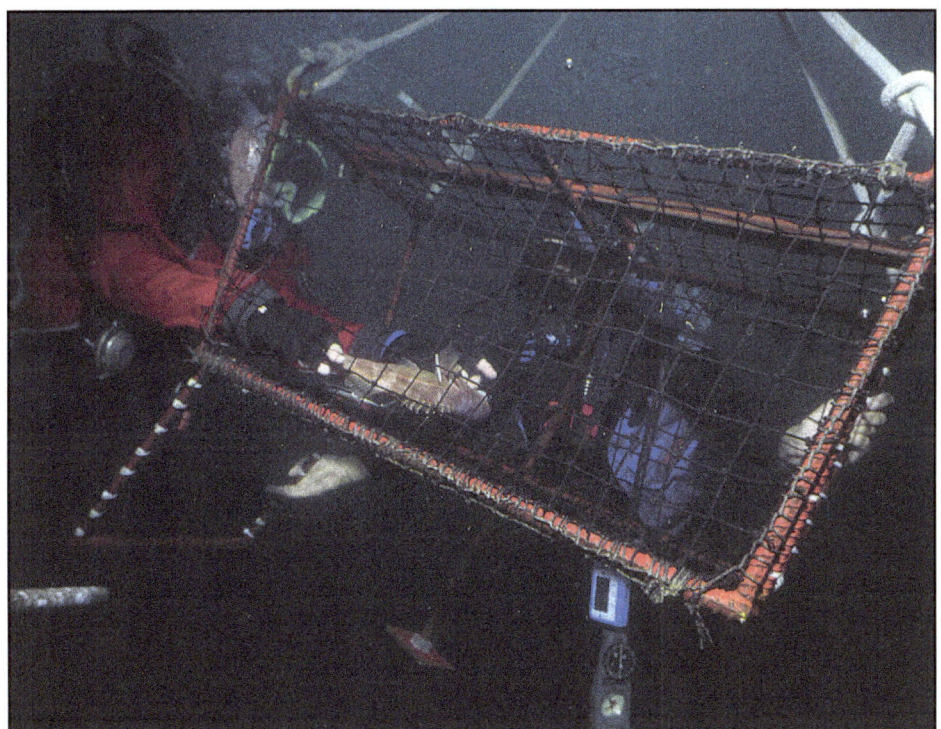

FIGURE 7.46:
Diver placing tagged fish into recovery cage for later release.
(photo courtesy of J. Heine)

1996). External tags are often embedded in the flesh of the fish, usually under the dorsal fin or through the head muscles. These will eventually work themselves out, and may last anywhere from days to years, depending upon the species. There are also a myriad of potential tagging effects that might adversely affect a fish or the study. These include tag loss, drag, infection, mortality, and attracting potential predators. Commonly used tags include Floy plastic anchor tags (Model FD-68 B), streamer tags (FTSL-73), garment tags, plastic disks (Floy FTF-69) (Thompson et al., 1986), wire, and monofilament line with various colored beads (Langley and Driscoll, 1989). Some of these tags can be custom ordered by color, with numbers, and with contact information for tag returns. A commonly-used tagging gun is made by Dennison (Mark II; available from Floy Tag). Fish can be caught by methods described above, and then tagged and released. Larger fish, such as some sharks, can be tagged *in situ* using a pole spear and strong stainless steel and monofilament Floy tag (model # FH-69) (Crane and Heine, 1993). These tags have lasted many years.

Fishes with swim bladders suffer a high degree of mortality when brought to the surface. Techniques have been developed to tag the fish underwater, at depth of ca. 20 m (60 ft) in order to avoid swim bladder expansion (Starr et al., 2002; Boland, 2005). Both visual and acoustic tags have been used.

Small coral reef fish, such as parrotfish and damselfish, have been tagged by subcutaneously injecting paint, dye, or elastomer (Thresher and Gronell, 1978; Hill and Grossman, 1987; Hixon and Beets, 1993). White acrylic paint can be used for dark colored fish, and concentrated blue histological stain can be used for lighter colored fish (Clifton, 1996). Blackburn et al., (2003) tagged red snapper with various colors of visual implant elastomer tags (Northwest Marine Technology, WA) between the caudal fin and the anal fin rays. They found that fluorescent colors were the easiest to distinguish using a dive light under low natural lighting. Other stains used include tattoo inks and liquid latex. Inks and dyes can be injected using stainless-steel dental needles inserted under several fish scales. Needles may need to be changed often to avoid clogging. Hundreds of unique combinations can be designed by using different locations on both sides of fish.

FIGURE 7.47:
Tagged damselfish with elastomer; note the orange tag on the body just below dorsal fin.
(photo courtesy of N. Schiel-Rolle)

Bridled gobies (*Coryphopterus glaucofraenum*) have been tagged with visible implant (VI) alpha tags, which are 1.0 mm x 2.5 mm long pieces of plastic film with a 3-digit alphanumeric code. The gobies are caught with hand nets and clove oil anesthetic (which slows them down). They are then sexed and measured with simple plastic dial calipers underwater. The tags are injected (while underwater) with a syringe with a flat needle, and placed just under the skin. The skin of this species is clear enough that the 3-digit code can be read

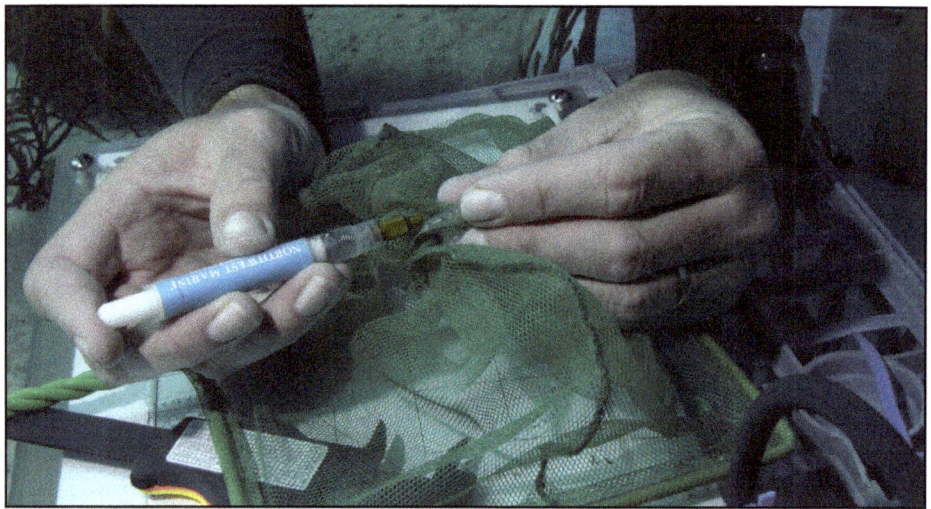

FIGURE 7.48:
Tagging gobies underwater with plastic film injected with a needle. Note calipers used to measure fish length.
(photo courtesy of M. Steele)

through the skin without having to recapture the fish.

Acoustic tags have been attached externally to sharks as a minimally intrusive procedure. An "acoustic bracelet tag" was designed to fall off due to a corrosive weak link after the battery was used up (Lobel, 2008). Sharks were captured by baited hook, tagged on the caudal peduncle, and then released.

Other methods of tagging include freeze-branding (Raymond, 1974; Berge, 1990), heat branding (Hargreaves, 1992), fin clipping (Sale, 1971), and marking of otoliths with tetracycline (Schmitt, 1984). See Coyer et al., (1999) for a thorough description of tagging and vendors for tags.

TRAINING EXERCISE #10
Collecting and Tagging Techniques

Suggested equipment: pole spear, suction gun, hand nets, goody bags, lobster hook, squeeze bottles, Ziploc bags, Floy tags and guns, cable ties, and other tagging items as available

Objective: Orientation and practice using collecting and tagging devices

The class can be divided into buddy pairs. A thorough briefing must be given by the teaching staff prior to the dive, which includes the dive plan and the use of each piece of equipment. Various stations can be set up close to each other, to allow the students to briefly visit each station and operate the equipment provided. It is helpful to have a staff member supervise each station.

Specific organisms can be targeted depending upon the collecting and tagging equipment available. Efforts should be made to expose the class to the widest variety of collecting and tagging gear as possible (see text), and the organisms should be released after capture. The class may be able to assist a local researcher with their collection or tagging project.

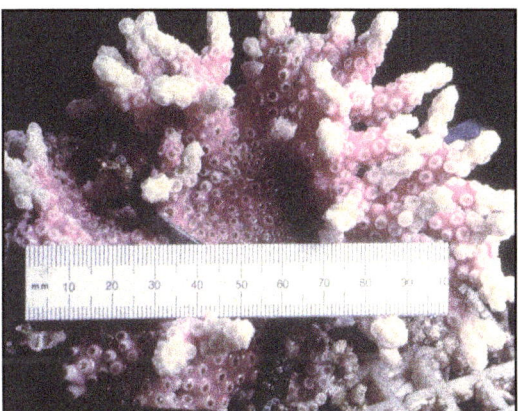

FIGURE 7.49:
Macroalgae and invertebrates can usually be measured with a straight ruler.
(photo courtesy of J. Reed)

Measuring and Size Frequency

SCIENTIFIC DIVERS frequently measure organisms to quantify growth or estimate size frequencies of populations. One of the simplest, yet often time-consuming methods, is to use a ruler to measure the total length of a plant, or the diameter of an invertebrate species, and record the size on a slate (Reed and Foster, 1984; Heine, 1989). A more accurate method for smaller organisms, such as test diameter in sea urchins, or carapace length in crabs, is the use of plastic or metal vernier calipers (Buttolph, 1989; Haaker et al., 2005). Shepherd (1985) describes an ingenious brass caliper gauge used to measure abalone. This gauge incorporates a strip of waterproof paper that slides along the gauge. The diver pushes on a thumb plate-backed needle, which punches a hole into the paper, allowing hundreds of measurements to be made in a short period of time. The paper can be easily changed underwater, and the measurements can be read later under a low power microscope or microfiche reader.

FIGURE 7.50:
More precise measurements, such as on this abalone, can be accurately made underwater using a caliper gauge.
(photo courtesy of K. Joe)

Mussel growth has been measured by notching the shell with a steel file at the growing edge to mark the initial shell length (Canestro et al., 1996). Plants, sponges, and other benthic organisms are often monitored for growth by periodic measurements or photographs of marked individuals (Dayton et al., 1974; Littler and Littler, 1985; Rutzler, 1996). Zieman

FIGURE 7.51:
A diver measures the length of a damselfish using a caliper gauge.
(photo courtesy of M. Steele)

(1974) marked seagrass blades using a modified stapler for growth studies. Other researchers have punched or bored holes into algal thalli and measured their movement over time (Mann and Mann, 1981). Branch elongation in calcareous algae was measured by using photography over time (Johansen and Austin, 1970).

Total length of fishes can also be estimated by divers. Bortone et al. (1992) estimated total length of fishes to the nearest 1 cm for fish less than 10 cm total length, and to the nearest 5 cm for fish larger than 10 cm total length. Length frequency histograms can be generated for an estimation of growth, recruitment, mortality, and population structure.

Accurate estimation of the size or length of fishes *in situ* can be difficult. There can be bias due to different observers, level of training or experience with the technique, water visibility, and fish mobility. One method of estimating fish length uses a ruler attached to the end of a one meter rod, which minimizes magnification problems in estimating fish size (Bohnsack and Bannerot, 1976). The only reliable method to ground-truth the visual estimation is to capture or spear the fish for actual measurement.

A paired-laser measuring device has been used as a training aid to reduce bias associated with estimation of fish sizes (McFall et al., 1992). They used commonly available red laser pointers mounted in housings fabricated from schedule 80 PVC. The lasers were mounted parallel to each other on a graduated 1 m aluminum bar, and one laser could slide along the bar to give a size measurement. Observers in a pool, using wood "targets," were able to estimate the size of the targets more accurately with the laser device than by observation alone, but the differences were only about 3% between the two methods.

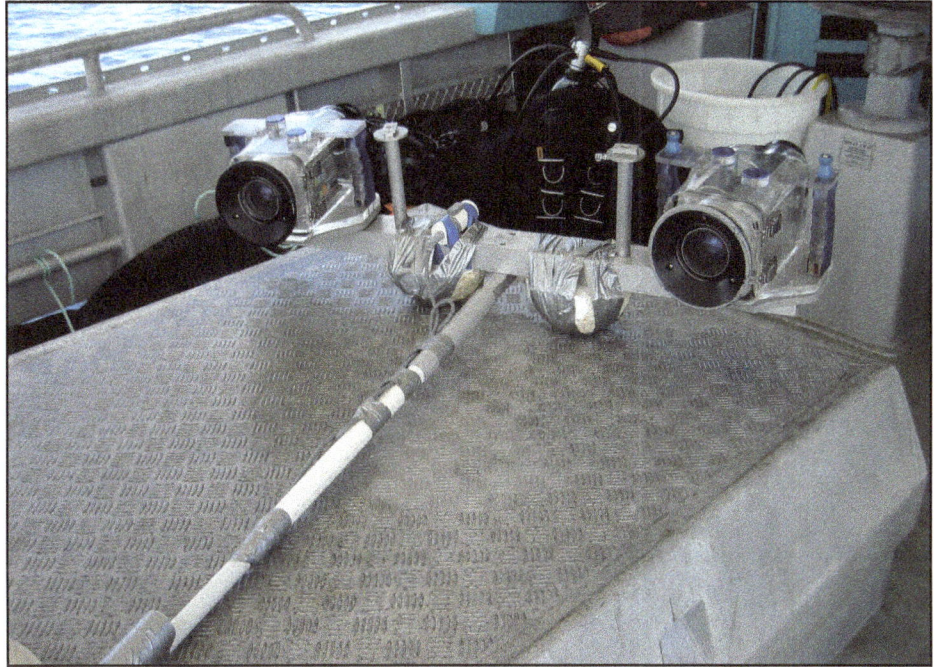

FIGURE 7.52:

Stereo-video cameras used to measure fish abundance and sizes.

(photo courtesy of E. Harvey)

FIGURE 7.53:
Large wire mesh cage enclosing a parrotfish.
(photo courtesy of M. Hay)

In the field, a pair of parallel lasers have been used to obtain information on the sizes of organisms (see Tusting and Davis, 1993, for complete description). Yoklavich et al. (1993) used lasers spaced 39.5 cm apart to estimate the size of rockfishes *in situ* and from video. When lasers are used in different numbers and arrangement, measurements can be made in both horizontal and vertical planes (Tusting and Davis, 1993).

Watson et al. (2010) used stereo-video camera images to measure fish lengths using the program PhotoMeasure (SeaGIS Pty Ltd). Measurements were only made on fishes within a maximum distance of 7 m from the cameras.

Caging

MANY TYPES OF ORGANISMS ARE EITHER HELD INSIDE or kept outside various enclosures for studies on water quality, behavior, ecology, growth, herbivory, and transplantation. The cages themselves can cause effects to the organisms being studied, so proper controls for "cage effects" may need to be considered (Virnstein, 1978; Hulberg and Oliver, 1980; Schmidt and Warner, 1984; Vadas, 1985; McMillan, 2008). Some problems that have been identified with cages and barriers include changes in light intensity, temperature, water flow, fouling, sedimentation, and attraction or protection of certain organisms (Hall et al., 1990).

Cages can be constructed from a variety of materials such as Vexar (Dupont) plastic mesh, wire fencing, construction material, or cotton, nylon, or polypropylene netting, stretched over a frame of PVC or steel pipe, rebar, or aluminum. Large cages (5 m diameter by 5 m high) constructed of netting with weights on the bottom and floats on the top, have been used to exclude predators for predation studies on newly settled fish (Hixon and Carr,

FIGURE 7.54:
Stainless steel mesh cages cemented with marine epoxy to the substrate, and marked with bicycle tape tag nailed into the rock.
(photo courtesy of M. Edwards)

1997). Hay (Aquarius mission 2010) used large wire mesh nets to enclose specific fish and mixes of fishes to look at the effects of grazing and herbivory on community structure.

Other types of barriers have been fabricated to keep various types of organisms inside or outside an area. Astroturf, plastic matting, sticky insect or rodent trap material, anti-fouling paint, wire mesh, fiberglass, and copper plating have all been used to prevent chitons, limpets, and sea urchins from moving across them (Foster and Sousa, 1985; Vadas, 1985). The upper edge of the fences can be bent inward or outward to control movement of grazers. In certain circumstances, just raising the experimental unit above the substrate will prevent herbivores from reaching it (Robles, 1982).

Copper paint will also repel many herbivores. McMillan (2008) used 1/2 inch copper pipe (schedule M) with 1/4 inch nylon mesh to surround giant kelp plants and exclude herbivorous gastropods, essentially creating a "moat" around the base of the plants. For a complete review of herbivory experiments, see Vadas (1985).

Transplantation and Outplantation
MICROSCOPIC ALGAL GAMETOPHYTES HAVE BEEN GROWN IN THE laboratory on glass slides and outplanted into the field (Hsiao and Druehl, 1973). *Fucus* eggs have been settled onto plastic strips and outplanted to the field (Pollock, 1969). Small specimens of newly recruited algae have been transplanted by tying small cobbles to lengths of chain with cable ties, by chipping rock with algae and cementing it in a new area, and by placing small algal thalli into the strands of braided line, and affixing it to the bottom (see Foster et al., 1985 for details). Large specimens of giant kelp have been transplanted by tying the holdfasts to sec-

FIGURE 7.55:
Outplanted boulders contained in a plastic tray.
(photo courtesy of M. Edwards)

tions of chain and towing them to a new site, where they are then affixed to earth anchors (Canestro et al., 1996), or by using bicycle inner tubes as large rubber bands to attach plants to rebar embedded in bags of cement. Individual rocks and boulders can be moved for experimental purposes (John and Pople, 1973).

Mussels are routinely transplanted and outplanted in the field in netted bags, and are used for a variety of analyses of water quality due to their filter-feeding mode (Rasmussen, 1996). Laboratory-raised abalone have been outplanted in small flow-through "bongo" cages constructed from short sections of PVC pipe with nitex netting on either end (Canestro et al., 1996). Sea urchins have also been outplanted and held in PVC-framed cages which are covered with Vexar (2 cm mesh size). Cages can be affixed to the bottom using underwater epoxy, threaded rods and nuts, or with weights or earth anchors. Other sessile organisms, such as encrusting algae, sea anemones, and cup corals have been removed from the substrate and reattached with underwater epoxy, and even attached with Velcro glued onto the organism (Paine, 1984).

FIGURE 7.56:
Diver affixing a coral using epoxy cement.
(photo courtesy of M. Miller)

FIGURE 7.57:
Gorgonians attached with epoxy to PVC plates and outplanted to the field.
(photo courtesy of M. Ponti)

Corals have been transplanted and outplanted to evaluate the performance of various lab-cultured and wild-collected specimens (Miller, 2010 Aquarius mission). Parameters

FIGURE 7.58:
Glass slides attached to fiberglass frames are often used for outplanting organisms that have been reared and settled in the laboratory. (photo courtesy of J. Heine)

FIGURE 7.59:
Wire mesh grazer exclusion cage with algal recruits on artificial substrate.
(photo courtesy of M. Edwards)

FIGURE 7.60:
Artificial substrata can be made of tiles and covered to exclude grazers.
(photo courtesy of K. Clifton)

measured include growth, survivorship, and surface microbial community composition.

Recruitment Techniques

THERE ARE A VARIETY of methods that have been employed to measure the recruitment of organisms. Many of these use either clearing of natural substrata or different types of artificial substrata such as settling plates (Neushul et al. 1976; Foster and Sousa, 1985; Airoldi and Cinelli, 1997). Giant kelp zoospores have been settled out onto glass slides which are then attached to fiberglass frames

FIGURE 7.61:
Recruitment cinder block hammered into substrate.
(photo courtesy of M. Hay)

underwater. The slides are collected at regular intervals for evaluation of sporophyte recruitment (Reed et al., 1994; Canestro et al., 1996).

Artificial substrata units that are protected from predators and grazers are used to measure early stages of community development. Rock (slate) settling plates, with cages around them, can be mounted in a variety of habitats and orientations to measure recruitment (Graham and Sebens, 1996). Flat PVC, concrete, Plexiglas, or ceramic tile have also been used to measure encrusting or fouling organism settlement (Foster and Sousa, 1985; Harriot and Fisk, 1987). However, there are potential problems with any artificial substrata with regard to differences from natural surfaces in color, chemical constitution, surface roughness, and rugosity (Foster and Sousa, 1985). Airoldi and Cinelli (1997) used Plexiglas panels affixed above the substrate to manipulate the sedimentation rates in recruitment studies. PVC pipe has been used as a simulated natural dead coral analog for recruitment studies with tropical encrusting coralline algae (Adey and Vassar, 1975). Polypropylene strips which mimic seagrass blades were used in epiphyte studies (Harlin, 1973). The placement of artificial substrata, with regard to positioning, angle, and distance off the bottom may all have an effect on recruitment and survivorship rates (Kennelly, 1983).

It is extremely difficult to remove all living material from rocky substrata underwater (Foster and Sousa, 1985). Many mechanical and chemical techniques have been used intertidally, but most are not appropriate underwater. Subtidally, tools such as knives, paint scrapers and wire brushes have been utilized. Exposing new rock surface with a hammer and chisel, or pneumatic tools, can also be done, but problems with leaching or weathering on newly exposed rock surfaces may affect recruitment (Dayton, 1971; Reed and Foster, 1984).

Chemical sterilization methods have been employed to remove all macro- and microscopic algal thalli from the substratum (Edwards, 1998). Plots were prepared for steriliza-

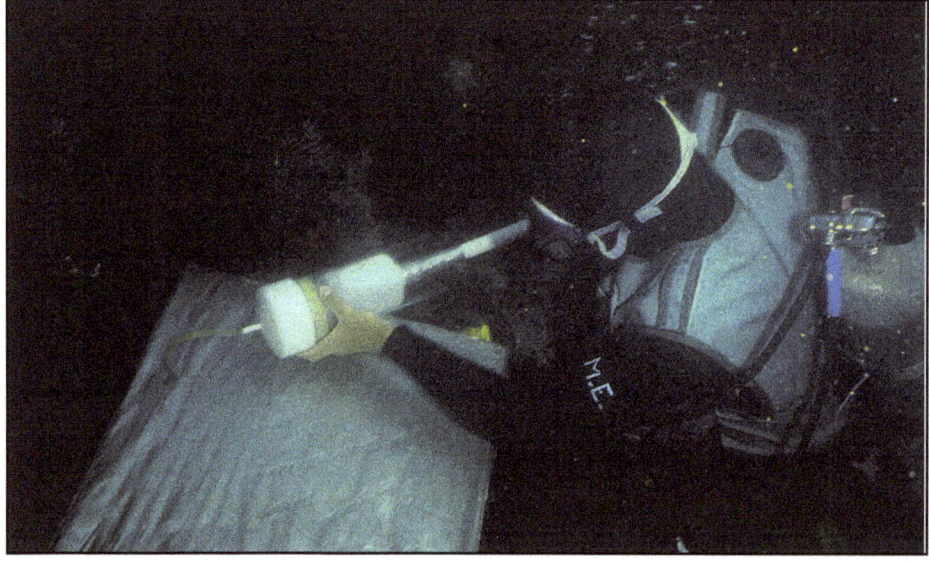

FIGURE 7.62:
Bleach is injected into a covered quadrat to sterilize the substratum for recruitment studies. (photo courtesy of M. Edwards)

tion by removing all macroscopic organisms (>0.1 cm) with a knife and then abrading the substrate with a wire brush. Two holes were drilled in the middle of two opposing sides of each plot with a pneumatic drill, and an 18 cm long stainless steel threaded rod was cemented into each hole. A 0.5 m^2 sterilization tent was placed over each plot and attached to the embedded threaded rods with stainless steel nuts. The tents, made from black plastic tarp material attached to 5 cm angle-stock PVC frames, were penetrated through the middle by a 16 cm piece of surgical tubing that served as a valve which could be opened or closed by simply untying or tying the tubing. A ~2.5 cm thick gasket made from modeling clay wrapped in thin plastic (to prevent it from dissolving) was attached to the base of each tent. Tightening the nuts on the threaded rod forced a water-tight seal between the tents and the substrate.

To check for leaks in the tent-substrate interfaces, 50 ml of Fluoricine dye (1 g Fluoricine / 50 ml filtered sea water) was injected into each tent with a hypodermic needle, and the tent perimeter was examined for leaking dye. Leaks were easily identified and plugged with a petroleum-base clay, making the tents water tight.

To remove all microscopic algal thalli from these experimental field plots, one liter of bleach was injected into each tent through its valve with a large syringe. The valves were then closed and the tents left for two days. Results showed that this method was more efficient than traditional scraping methods.

Artificial Reefs

A WIDE VARIETY OF ARTIFICIAL REEFS AND UNDERWATER habitats have been designed for increasing populations of organisms impacted due to overfishing, or as mitigation for environmental damage. In Florida alone, 15 reefs were constructed prior to 1970, 92 reefs in the 1970s, and 173 reefs by the early 1980s (Halusky, 1986). By the year 2009, 2,267 articicial reefs have been constructed (http://edis.ifas.ufl.edu/fe649). The design of such reefs must take into account the target species, and match their habitat requirements to the materials and technology available, as well as considerations for durability in harsh weather conditions (Spanier et al., 1985).

Triangular-shaped reefs constructed of used automobile tires held together with steel and fiberglass and anchored to the

FIGURE 7.63:
A rockfish inhabits an artificial reef constructed of cinder blocks. (photo courtesy of J. Heine)

FIGURE 7.64:
Artificial reefs used to study goby reproduction.
(photo courtesy of S. Painter)

bottom with concrete have been used in the Mediterranean Sea (Spanier et al., 1985). A large artificial reef system was built in southern California using quarry rock, which was dumped to form individual modules of varying configurations and dimensions (Aseltine et al., 1985).

A benthic structure designed for the purpose of growing mussels and other mollusks is described by Hampson et al. (1989). Their structure was designed to sit off the sea floor, rather than suspend from the surface, and to keep the organisms off the bottom. An upper air-filled PVC pipe provides buoyancy for grow-out lines that are attached to the rig.

Aspects of goby population structure, such as density and sex ratio, were manipulated using "artificial reefs" constructed of PVC pipe and Vexar mesh (see Figure 7.64; S. Painter). Rocks were moved into the reefs using 5 gallon buckets and lift bags. Each reef has six buckets of rocks. The rocks were placed on top of a one meter square piece of plastic fencing to slow the rate at which they sink into the sand. The cage was secured to the bottom with reusable cable ties so that divers can easily open and close the cage. To monitor reproduction, several artificial nesting sites were added to each reef. The artificial nests are a small piece of PVC pipe with a cap at one end. Inside the nest is a clear piece of acetate paper rolled up. The fish lay their eggs on the acetate paper which can be easily removed and photographed by divers.

Artificial reefs constructed of concrete blocks on plywood foundations with wire mesh to prevent fish from escaping have been used to study predation, prey refuges, and the

FIGURE 7.65:
Porcupine fish attractor device made of PVC.
(photo courtesy of S. Halewood)

structure of coral-reef fish assemblages (Hixon and Beets, 1993).

Fish aggregating devices (FADs) are physical structures placed in areas that are flat topographically, or in mid-water, and relatively devoid of fishes. Structures such as sunken vessels, steel modules, surplus concrete pipe, plastic cones and hemispheres, abandoned offshore platforms, and transplanted kelp have all been used to attract fish (Potts and Hulbert, 1992). Two types of mid-water FADs are described by Potts and Hulbert (1992). Three-dimensional structures attracted significantly higher numbers of fishes than did smaller, two-dimensional structures. "Porcupine Fish Attractors" were used to study deep-water recruitment of rockfishes off Santa Barbara (Love, unpubl.). Bohnsack (1991) gives a good description of habitat structure and the design of artificial reefs. Shelter characteristics of artificial reefs for Caribbean fish are described by Hixon and Beets (1989).

Telemetry

MOVEMENTS OF MARINE ORGANISMS CAN NOW be tracked using sophisticated telemetry methods. This can give information on such things as habitat utilization, feeding behavior, and the ecology of organisms. Electronic components have been reduced in size to the point where even very small organisms can now be tagged and tracked. Environmental parameters such as temperature, depth, and time can also be recorded.

Initially, simple position-finding beacons were used. Now, systems have been used to monitor movements of fish (Carr and Chaney, 1976; Stasko and Pincock, 1977; Winter,

1983; Starr et al., 2002; Lindholm et al., 2009), lobsters, seals, and birds, and even to measure physiological functions (Wolcott and Hines, 1996). Phenomena such as EKG, temperature, posture, sound, light, depth, swimming speed, and orientation can now be measured *in situ*.

Blue crab habitat use has been studied by using telemetry because the water is too murky for direct observations (Wolcott and Hines, 1996). Custom made transmitters about 2 cm long were attached to the crab's carapace, and electrodes inserted into the mandible muscles detected "chewing". Magnetic reed switches and magnets were placed on the crabs "elbows", to measure agonistic behavior. They also were able to track pre-molt and post-molt crabs to determine preferred habitats during those time periods.

Acoustic transmitters were surgically implanted in rockfish *in situ* in mid-water by scientific divers (Starr et al., 2002). In order to avoid gas bladder expansion and thermal shock, deep water rockfish were brought up on a long line to a depth of 20 m (60 ft), anesthetized with quinaldine, and strapped into a surgery center. The surgeon would scrape off scales in the abdomen, make a small incision, and insert a radio transmitter. Betadine was squirted into the incision to minimize infection, and the incision was closed using surgical staples. The fish was transferred to a recovery cage, which had doors that opened upon contact with the bottom, thus freeing the fish near where it was caught. Fish were tracked using fixed VEMCO (Shad Bay, Nova Scotia) acoustic receiver arrays. The effect of topographic relief on transmitter signal detection must be carefully evaluated (Lindholm et al., 2009).

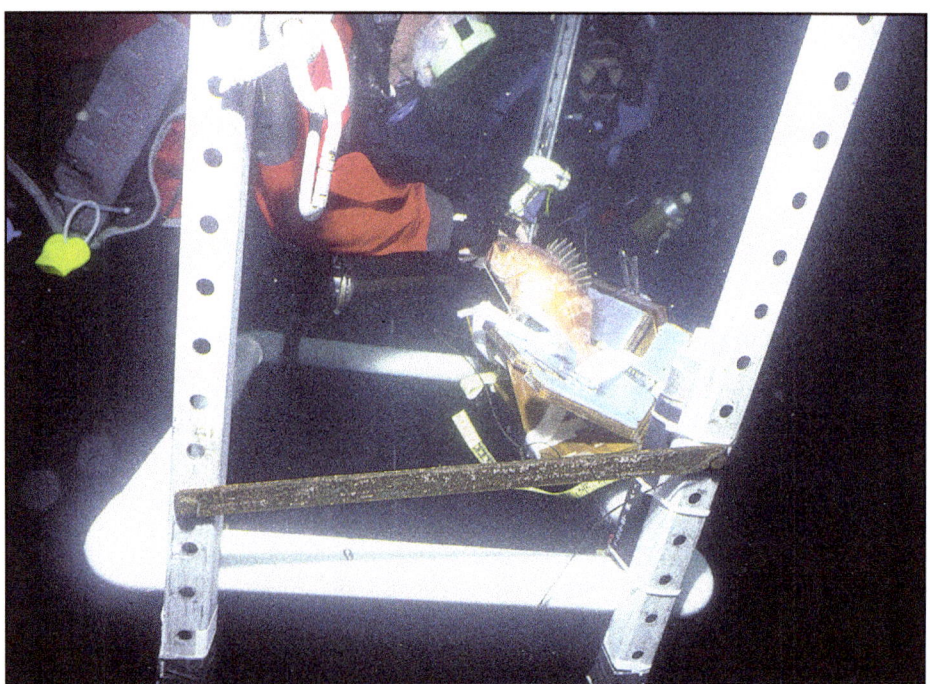

FIGURE 7.66:
Diver operating on a rockfish in a mid-water "surgery center" to implant acoustic transmitters. (photo courtesy of J. Heine)

FIGURE 7.67:
Direct observations by divers using rebreathers often uncover behaviors that might otherwise not be seen.
(photo courtesy of D. Kesling)

Behavior

COMPLEX BEHAVIORAL INTERACTIONS, SUCH as reproduction in pelagic spawning fishes, can be analyzed by using video records played at slow, or single-frame speeds, or by direct diver observations. Quantifiable spawning behaviors in the bluehead wrasse (*Thalassoma bifasciatum*) included spawning rushes, which are rapid upward swims that culminated in a visible gamete cloud, possible rushes where a gamete cloud could not be detected, and false rushes in which the upward swim was interrupted before gametes could be released (Clavijo and Lindquist, 1992). Juvenile rockfish behavior has been monitored *in situ* by direct observation, and by noting reaction to a fish's image in a mirror (Hoelzer, 1982). Predatory behavior in four piscivorous reef fish species was observed *in situ* and recorded on video by Auster (2005). Hixon (1991) gives a thorough review of predation as a process that structures coral reef communities.

Nocturnal and crepuscular behavior of reef fishes have also been evaluated using direct observation and video, as well as with an ROV (Lindquist and Clavijo, 1992). Using a modification of the stationary survey method developed by Bohnsack and Bannerot (1986), with a 4 m (13 ft.) radius circle of observation, they made day and night counts of fishes and observations of fish feeding behavior and changing color patterns at night. The authors noted that harsh white lights on the video camera tended to "disturb" fishes, so they used red filters over the lights, but still noticed some avoidance behavior by the fishes. Lights also inhibited some feeding behaviors in the nocturnal fishes they observed.

Swimming behaviors of scyphomedusae have been recorded on video in the field (Klos et al., 1996). Using a Hi-8mm video camera in a waterproof housing, the researchers recorded jellyfish swimming near the surface, in the water column, and near the sea floor, under natural lighting. The videotape was analyzed for time budgets of swimming and resting periods.

Coral larval swimming behavior has been studied in clear acrylic tubes (1 cm diameter, 50 cm tall) suspended from buoy lines. Using a flashlight and a white slate background allows for viewing of the 1 mm coral larvae during day or night (see Canestro, 1997 for brief discussion).

A detailed discussion of observations of certain marine animal behavior, including pelagic fish, groundfish, shrimp, and octopuses is given by High (1998). Many accounts are anecdotal, but they provide interesting observations of some rarely seen organisms.

Field observations by divers have shown that dusky damselfish demonstrate agnostic guarding behavior and selective algal farming (Di Santo et al., 2009). Territory size, algal garden composition, and encounters with other fish were recorded using a Sony Handycam DCR-SR40 in a waterproof housing.

The behavior of most pelagic organisms would not be known if it were not for blue-water diving techniques (see Chapter 3 and Haddock and Heine, 2005). Discoveries include larvacean houses (Alldredge, 1977), associations between gelatinous zooplankton and hyperiid amphipods (Gasca and Haddock, 2004), feeding behavior of salps (Madin, 1974), foraging, feeding, and escape behavior of ctenophores (Matsumoto and Harbison, 1993; Haddock, 2007), feeding in the scyphozoan medusae *Aurelia* (Klos et al., 2005), and krill behavior (Ragulin, 1969).

Trawl Diving

FISHERY BIOLOGISTS MAY DIVE NEXT TO OR HOLDING onto various types of trawls in order to evaluate their performance underwater. This can be an extremely demanding and potentially dangerous type of diving, often at speeds of up to 5 knots (Workman, 1986). Specialized equipment used includes adjustable second stage regulators to decrease the tendency to free-flow, a small knife worn on the forearm for use in entanglements, diver to diver and to surface communications, and a diving sled.

Diving techniques and safety on trawls have been described by a number of researchers (High 1967; Main and Sangster, 1978; Wickham and Watson 1976; Workman, 1986). Scientific divers might do things such as measure flow rates around the trawl, measure spread or openings or door angles in trawls, observe animal behavior, and photograph or videograph the trawl (High, 1998).

Scientists from the Bureau of Commercial Fisheries and National Marine Fisheries Service conducted a number of early dives around fishing gear (High, 1998). They pioneered the use of sea sleds and free-swimming techniques to observe mid-water and bottom trawls *in situ*. They also dived in purse seines containing yellowfin tuna, dolphins, and sharks, and around gill nets, fish traps, and long lines, developing special diving techniques for these potentially dangerous situations.

FIGURE 7.68:
Divers putting a temperature probe on a dolphin.
(photo courtesy of J. Estes)

Working with Birds and Mammals Underwater

MANY VERTEBRATES, ESPECIALLY marine mammals, require permits to handle and even approach within a certain distance. In order to hold marine mammals in captivity in the U.S., permits are required from NOAA and the Animal Plant Health Inspection Service (APHIS). Additional university authorization may also be required. There is an organized group of animal trainers called the International Marine Animal Trainers Association (IMATA) that holds annual conferences and publishes a magazine called *Soundings*.

The U.S. Navy has pioneered the training methods with marine mammals. More recently, aquariums and zoos with marine mammal displays do considerable training. They also have divers, often on surface supply, cleaning tanks and displays while marine mammals are present.

For scientific divers in the water with marine mammals, there are certain basic safety considerations that should be followed. Divers

FIGURE 7.69:
Divers training a California sea lion who is wearing an underwater video camera on its back.
(photo courtesy of J. Rotman)

FIGURE 7.70:
Divers must approach large fish, such as this whaleshark, and mammals with caution.
(photo courtesy of M. Awai)

need to always be aware of their surroundings and not jeopardize their safety at the expense of an animal under training. Seals and sea lions can safely make much more rapid ascents than scuba divers! Divers also should be sure that their equipment is streamlined, and that no extraneous hoses or other equipment are dangling away from the diver that might distract the animal, or give it the opportunity to bite the equipment. If the animal is on a tether, divers must be careful not to become entangled, and should carry a sharp knife or safety shears.

Scientists at Moss Landing Marine Laboratories are training California sea lions to carry a video camera to film gray whales and other pelagic species in the wild. Dolphins have been videotaped for motion analysis, and voltmeters in underwater housings have been used to measure heat flow and skin temperature with a thermistor (Canestro, 1997).

Other important work with marine birds and mammals can be found in (Kooyman, 1966; Kooyman et al., 1971; Ridgway, 1972; and Costa, 1991).

Diving with Captive Fish and Mammals

MANY AQUARIUMS HAVE DIVING PROGRAMS FOR scientific research, educational, husbandry, and maintenance reasons. Some also allow the public to dive in certain exhibits. There are a number of special circumstances and conditions that must be evaluated, such as animal welfare issues, fragility of exhibits, contamination issues, and potentially dangerous animals. Diving with captive sharks is an example of a situation that requires a risk assessment for the health and safety of both the sharks and the divers (Hodges and Frierson, 2008). A survey of many aquariums in the U.S. and abroad found that there is a wide variety

FIGURE 7.72:
Aquarium diver in a shite shark net pen acting as a safety diver. (photo courtesy of G. Peterson)

FIGURE 7.71:
Aquarium diver observing large manta ray.
(photo courtesy of M. Awai)

of shark species on display, and a number of safety methods employed. Safety methods included separation by nets or physical barriers, chain mail suits, electronic repellents (e.g. Shark Shield), and the use of sticks.

The AAUS has established standards for scientific aquarium diving that allow for certain departures from traditional buddy system diving due to the unique nature of this activity (AAUS, 2009).

MANUFACTURERS AND SUPPLIERS

Scientific equipment and underwater tools

Aquatic Research Instruments (plankton nets, water samplers, sediment corers)
P.O. Box 98, 620 Wellington Pl.
Hope, ID 83836
(800) 320-9482
www.aquaticresearch.com

The Bee Works (bee tags with glue)
5 Edith Drive
Orillia, ON, CANADA
L3V 6H2
www.beeworks.com
http://www.beeworks.com/catalog/index.php?main_page=product_info&cPath=10&products_id=125

Floy Tag, Inc. (wide variety of tags)
4616 Union Bay Pl. N.E.
Seattle, WA 98105
(800) 843-1172
www.floytag.com

Forestry Suppliers (transect tapes, compasses, inclinometers; DO and pH meters)
205 W. Rankin St. Box 8397
Jackson, MS 39284
(800) 752-8460
www.forestry-suppliers.com

Heinz Walz (Dive PAM fluorometer, light meters)
Eichenring 6, 91090 Effeltrich
Germany
www.walz.com

JBL Enterprises, Inc. (spearguns)
www.lblspearguns.com

Memphis Net & Twine Co. (nets, line, floats, seines, traps)
P.O. Box 80331
Memphis, TN 38108
(901) 458-2656
(888) 674-7638
www.memphisnet.net

National Band and Tag Co. (Tags)
Box 72430
Newport, KY 41072
(859) 261-2035
www.nationalband.com

Nalge Co. Thermo Fisher Scientific (Nalgene waterproof paper; PolyPaper Plastic Paper)
75 Panorama Creek Dr.
Rochester, NY 14625
(800) 625-4327
www.nalgenelabware.com

Nichols Net and Twine Co. (netting, fish traps)
(800) 878-6387
http://www.nicholsnetandtwine.com/

Outer Banks Outfitters (marine supplies, GPS)
P.O. Box 3330
Atlantic Beach, NC 28512
(877) 690-0007
http://www.outerbanksoutfitters.com/

Overton's (boating accessories)
111 Red Banks Road
P.O. Box 8228
Greenville, NC 27835
(800) 334-6541
www.overtons.com

Progressive Epoxy Polymers (marine epoxy, resins)
http://www.epoxyproducts.com/

Ramset Fastening Systems (drills, hammers)
3405 Dallas Hwy SW
Bldg 800 Ste. 810
Marietta, GA 30064
(800) 241-5640
www.ramset.com

Rite in the Rain (waterproof paper)
(253) 922-5000
www.riteintherain.com

The EarTag Company (animal ear tags)
CCK Outfitters
P.O. Box 650
Aubrey, TX 76227
(800) 977-4225
http://www.cckoutfitters.com/EARTAGS.html

VEMCO (acoustic telemetry equipment)
100 Osprey Dr.
Shad Bay
Nova Scotia
Canada B3T 2C1
(902) 852-3047
www.vemco.com

West Marine (marine supplies)
(800) 262-8464
www.westmarine.com

YSI (instruments, sensors, water quality)
(800) 765-4974
www.ysi.com

Websites

http://www.epoxyproducts.com/uwhistory.html (how-to for underwater epoxy)

http://www.conwedplastics.com/products.asp (webbing, mesh, netting, caging material)

http://www.redden-net.com/vexar.cfm (vexar, netting, seines, gillnets, trawls)

http://www.reefcheck.org/rcca/rcca_home.php (Reef Check California)

http://reefcheck.org/rcca/monitoring_protocol.php (Reef Check California monitoring protocol)

http://www.nightsea.com/lightson.htm (fluorescence)

http://www.mesophotic.com/ (mesophitic coral ecosystems)

References

Adey, W.H. and J.M. Vassar. 1975. Colonization, succession and growth rates of tropical crustose coralline algae (Rhodophyta, Cryptonemiales). Phycologia 14: 55-69.

Airoldi, L. and F. Cinelli. 1997. Effects of sedimentation on subtidal macroalgal assemblages: an experimental study from a Mediterranean rocky shore. J. Exp. Mar. Biol. Ecol. 215: 269-288.

Alcock, J. 1993. Animal behavior: An evolutionary approach. Sinauer, 625 pp.

Aleem, A.A. 1973. Ecology of a giant kelp bed in southern California. Bot. Mar. 16: 83-95.

Alldredge, A.L. 1977. Abandoned larvacean houses: a unique food source in the pelagic environment. Science 177: 885-887.

Alldredge, A.L. and M.W. Silver. 1988. Characteristics, dynamics and significance of marine snow. Prog. Oceanogr. 20: 41-82.

Aloi, J. E., S. L. Loeb, S. H. Hackley, C. R. Goldman, and A. T. Aloi. 1985. Underwater research methodologies for limnological investigations at Lake Tahoe, California-Nevada: periphyton, groundwater and zooplankton studies. In: Mitchell, C.T. (ed.), Diving for Science...85, Proceedings of Joint International Scientific Diving Symposium, Amer. Acad. Underwater Sci. pp. 313-317.

Ambrose, R.F., J.M. Engle, J.A. Coyer, and B.V. Nelson. 1993. Changes in urchin and kelp densities at Anacapa Island, California. 3rd California Islands Symposium, Nat. Park Serv. http://hdl.handle.net/10139/3178.

Anderson, T.W. and M.H. Carr. 1997. BINCKE: A highly efficient net for collecting reef-associated fishes. Envir. Biol. Fishes.

Andrew, N.L. and B.D. Mapstone. 1987. Sampling and the description of spatial pattern in marine ecology. Oceanogr. Mar. Biol. Ann. Rev. 25: 39-90.

Aseltine, D. A., R. D. Lewis, K. C. Wilson, and H. A. Togstad. 1985. Epibenthic community development on a quarry rock artificial reef. In: Mitchell, C.T. (ed.), Diving for Science...85, Proceedings of Joint International Scientific Diving Symposium, Amer. Acad. Underwater Sci. pp. 136-149.

Auerbach, P.S. 1997. A medical guide to hazardous marine life. Best Publishing, Flagstaff, AZ.

Auster, P.J. 2005. Predatory behavior of piscivorous reef fishes varies with changes in landscape attributes and social context: Integrating natural history observations in a conceptual model. In: Godfrey, J.M. and S.E. Shumway (eds.), Diving for Science 2005, Proceedings of Amer. Acad. Underwater Sci. pp. 115-127.

Balcom, P.H., J.M. Godfrey, D.C. Bennett, G.A. Grenier, C.G. Cooper, D.R. Cohen, D.A. Arbige, and W.F. Fitzgerald. 2007. Deploying benthic chambers to measure sediment oxygen demand in Long Island Sound. In: Pollock, N.W. and J.M. Godfrey (eds.), Diving for Science 2007, Proceedings of Amer. Acad. of Underwater Sci. pp. 135-141.

Baldwin, C.C., B.B. Collette, L.R. Parenti, D.G. Smith, and V.G. Springer. 1996. Collecting fishes. In: Lang, M.A. and C.C. Baldwin (eds.), Methods and Techniques of Underwater Research, Proceedings of Amer. Acad. of Underwater Sci. pp. 11-33.

Barilotti, D.C. and W. Silverthorne. 1972. A resource management study of *Gelidium robustum*. In: Nisizawa, K. (ed.), Proc. Seventh International Seaweed Symp., Univ. Tokyo Press, Tokyo. pp. 255-261.

Barry, J. and R. Coggan. 2010. The visual fast count method: Critical examination and development for underwater video sampling. Aquat. Biol. 11: 101-112.

Barsky, S.M. 1990. Diving in high risk environments. Dive Rescue Inc. International, Fort Collins, CO. 118 pp.

Barsky, S.M. 1997. Spearfishing for skin and scuba divers. Best Publishing, Flagstaff, AZ.

Bell, J.D., G.J.S. Craik, D.A. Pollard, and B.C. Russell. 1985. Estimating length-frequency distributions of large reef fish underwater. Coral Reefs 4: 41-44.

Berge, G.M. 1990. Freeze branding of Atlantic halibut. Aquacult. 89: 383-386.

Biggs, D.C., P. Laval, J-C. Braconnot, C. Carre, J. Goy, M. Masson, and P. Morand. 1986. In: Mitchell, C.T. (ed.), Diving for Science...86, Proceedings of Amer. Acad. Underwater Sci. pp. 153-161.

Birkeland, C.T. 1974. Interactions between a seastar and seven of its predators. Ecol. Monogr. 44: 211-232.

Bishop, M.A., B.F. Reynolds, and S.P. Powers. 2010. An *in situ*, individual-based approach to quantify connectivity of marine fish: Ontogenetic movements and residency of lingcod. PLoS ONE 5(12): e14267. Doui: 10.1371/journal.pone.0014267.

Blackburn, B.R., N. Brennan, and K. Leber. 2003. *In situ* diver identification of hatchery released red snapper, *Lutjanus campechanus*, using visual implant elastomer tags in the Gulf of Mexico (abstract only). In: Norton, S.F. (ed.), Diving for Science...2003, Proceedings of Amer. Acad. Underwater Sci. pp. 19.

Blumer, L.S. 1984. Simple, inexpensive method of tagging ictalurid fishes for individual identification. Prog. Fish. Cult. 46: 152-154.

Bohnsack, J.A. 1991. Habitat structure and the design of artificial reefs. In: Bell, S.S., E.D. McCoy, and H.R. Muchinsky (eds.), Habitat structure: The physical arrangement of objects in space, pp. 412-426.

Bohnsack, J.A. and S.P. Bannerot. 1982. A random point census technique for visually assessing coral reef fishes. The visual assessment of fish populations in the Southeastern United States: 1982 Workshop. South Carolina Sea Grant Technical Report, 1982; 1: 5-7.

Bohnsack, J.A. and S.P. Bannerot. 1986. A stationary visual census technique for quantitatively assessing community structure of coral reef fishes. NOAA Tech. Rept. NMFS 41: 1-15.

Boland, R. 2005. The Pacific Islands Fisheries Science Center dive program: meeting the challenges of the Pacific region. In: Godfrey, J.M. and S.E. Shumway (eds.), Diving for Science 2005, Proceedings of Amer. Acad. Underwater Sci. pp. 179-189.

Boland, R.C. and M.J. Donohu. 2003. Marine debris accumulation in the nearshore marine habitat of the endangered Hawaiian monk seal, *Monachus schauinslandi*, 1999-2001. Mar. Poll. Bull. 46: 1385-1394.

Bortone, S.A., J.V. Tasell, A. Brito, J.M. Falcon, and C.M. Bundrick. 1992. Visual census as a means to estimate standing biomass, length, and growth in fiches. In: Cahoon, L.B. (ed.), Diving for Science...1992, Proceedings of Amer. Acad. Underwater Sci. pp. 13-21.

Breda, V.A. 1982. Composition, abundance and phenology of foliose red algae associated with two *Macrocystis pyrifera* forests in Monterey Bay, California. Master's thesis, San Jose State Univ., CA 67 pp.

Brock, R. E. 1982. A critique of the visual census method for assessing coral reef fish populations. Bull. Mar. Sci. 32: 269-276.

Brock, V. E. 1954. A preliminary report on a method of estimating reef fish populations. Journ. Wild. Mgt. 18: 297-308.

Brook, I.M. 1979. A portable suction dredge for quantitative sampling in difficult substrates. Estuaries 2(1): 54-58.

Buggeln, R.G. 1985. Carbon allocation, pp. 415-425. In: Littler, M.M. and D.S. Littler (eds.). Handbook of Phycological Methods. Ecological Field Methods: Macroalgae. Cambridge University Press. 617 pp.

Buggeln, R.G. and S. Lucken. 1979. Kinnetic characteristics of photoassimilate translocation in *Alaria esculenta* (Laminariales, Phaeophyta). Planta 147: 214-245.

Butman, C.A. 1986. Sediment trap braces in turbulent flow: results from a laboratory flume study. J. Mar. Res. 44: 645-693.

Buttolph, P. 1989. Red urchin survey of northern California. In: Lang, M.A. and W.C. Jaap, (eds.), Diving for Science...1989, Proceedings of Amer. Acad. Underwater Sci. pp. 29-37.

Cahoon, L.B., D.G. Lindquist, I.E. Clavijo, and C. R. Tronzo. 1992. Sampling small invertebrates at the sediment-water interface. In: Cahoon, L.B. (ed.), Diving for Science...1992, Proceedings of Amer. Acad. Underwater Sci. pp. 51-59.

Cahoon, L.B. 1996. An instrumented lander for measurement of benthic respiration and production. In: Lang, M.A. and C.C. Baldwin (eds.), Methods and Techniques of Underwater Research, Proceedings of Amer. Acad. Underwater Sci. pp. 45-51.

Campbell, D.B. 1984. Foraging movements of the sea star *Asterias forbesi* in Narragansett Bay, Rhode Island, USA. Mar. Behav. Physiol. 11: 185-198.

Campbell, R.M. and D.H.N. Spence. 1976. Preliminary studies on the primary productivity of macrophytes in Scottish freshwater lochs. In: Drew, E.A., J.N. Lythgoe, and J.D. Woods, (eds.), Underwater Research. 1976. Academic Press, London. pp. 347-355.

Canestro, D., P.T. Raimondi, D.C. Reed, R.J. Schmitt, and S.J. Hollbrook. 1996. A study of methods and techniques for detecting ecological impacts. In: Lang, M.A. and C.C. Baldwin (eds.), Methods and Techniques of Underwater Research, Proceedings of Amer. Acad. Underwater Sci. pp. 53-60.

Canestro, D. 1997. Scientific diving methods and techniques slug (UCSC) style. In: Maney, Jr. E.J. and C.H. Ellis, Jr. (eds.), Diving for Science...1997, Proceedings of Amer. Acad. Underwater Sci. pp. 25-27.

Carleton, J.H. and P.W. Sammarco. 1987. Effects of substratum irregularity on success of coral settlement: quantification by comparative geomorphological techniques. Bull. Mar. Sci. 40: 85-98.

Carr, W.E.S. and T.B. Chaney. 1976. Harness for attachment of an ultrasonic transmitter to the red drum, *Sciaenops ocellata*. Fish. Bull. U.S. 74: 998-999.

Carpenter, R.C. 1984. Predator and population density control of homing behavior in the Caribbean echinoid *Diadema antillarum*. Mar. Biol. 82: 101-108.

Chapman, A.R.O. 1985. Demography, pp. 251-268. In: Littler, M.M. and D.S. Littler (eds.). Handbook of Phycological Methods. Ecological Field Methods: Macroalgae. Cambridge University Press. 617 pp.

Cheal, A.J. and A.A. Thompson. 1997. Comparing visual counts of coral reef fish: Implications of transect widths and species selection. Mar. Ecol. Prog. Ser. 158(1-3): 241-248.

Chenelot, H., S. Jewett, and M. Hoberg. 2008. Invertebrate communities associated with various substrates in the nearshore eastern Aleutian Islands, with emphasis on thick crustose coralline algae. In: Brueggeman, P. and N.W. Pollock (eds.), Diving for Science 2008, Proceedings of Amer. Acad. Underwater Sci. pp. 13-36.

Chess, J.R. 1978. An airlift sampling device for *in situ* collecting of biota from rocky substrata. MTS Journal 12: 20-23.

Chiappone, M., K.M. Sullivan, and J. Tschirky. 1996. Coral-reef assessment and monitoring methods: Examples from the Florida Keys and Caribbean. In: Lang, M.A. and C.C. Baldwin (eds.), Methods and Techniques of Underwater Research, Proceedings of Amer. Acad. Underwater Sci. pp. 61-74.

Christie, M.R., D.W. Johnson, C.D. Stallings, and M.A. Hixon. 2010. Self-recruitment and sweepstakes reproduction amid extensive gene flow in a coral-reef fish. Mol. Ecol. 19(5): 1042-1057.

Clavijo, I.E. and D.G. Lindquist. 1992. Quantifying pelagic spawning rushes in labroid fishes: Preliminary comparison of direct diver and video count techniques. In: Cahoon, L.B. (ed.), Diving for Science...1992, Proceedings of Amer. Acad. Underwater Sci. pp. 73-76.

Clavijo, I.E., D.G. Lindquist, S.K. Bolden, and S.W. Burk. 1989. Diver inventory of a mid-shelf reef fish community in Onslow Bay, NC: Preliminary results for 1988 and 1989. In: Lang, M.A. and W.C. Jaap, (eds.), Diving for Science...1989, Proceedings of Amer. Acad. Underwater Sci. pp. 59-65.

Clifton, K.E. 1996. Field methods for the behavioral study of foraging ecology and life history of herbivorous coral-reef fishes. In: Lang, M.A. and C.C. Baldwin (eds.), Methods and Techniques of Underwater Research, Proceedings of Amer. Acad. Underwater Sci. pp. 61-74.

Colwell, R.R. (ed.). 1982. Microbial hazards of diving in polluted waters. Maryland Sea Grant Pub. No. UM-SG-TS-82-01.

Cooper, R. 1970. Retention of marks and their effects on growth, behavior, and immigrations of the American lobster, *Homarus americanus*. Trans. Amer. Fish. Soc. 99: 409-417.

Costa, D.P. 1991. Reproductive and foraging energetics of high latitude penguins, albatrosses and pinnipeds: Implications for life history patterns. Amer. Zool. 31: 111-130.

Coyer, J., and J. Witman. 1990. The Underwater Catalog: A Guide to Methods in Underwater Research. Shoals Marine Laboratory, Cornell University, NY. 72 pp.

Coyer, J., D. Steller and J. Witman. 1999. The Underwater Catalog: A guide to methods in underwater research. 2nd Edition, Shoals Marine Laboratory, Cornell Univ., Ithaca, NY. 151 pp.

Crane, N.L. and J.N. Heine. 1993. Observations of the prickly shark (*Echinorhinus cookei*) in Monterey Bay, California. Calif. Fish and Game 78(4): 166-168.

Dahl, A. 1973. Surface area in ecological analysis: Quantification of benthic coral reef algae. Mar. Biol. 23: 239-249.

Dance, C. 1987. Patterns of activity of the sea urchin *Paracentrotus lividus* in the Bay of Port-Cros (Var, France, Mediterranean). Mar. Ecol. 8: 131-142.

Davies, I.J. and D.J. Ramsey. 1989. A diver-operated suction gun and collection bucket for sampling crayfish and other aquatic macroinvertebrates. Can. J. Fish. Aquat. Sci. 46: 923-927.

Davis, G.E. and T.W. Anderson. 1989. Population estimates of four kelp forest fishes and an evaluation of three *in situ* assessment techniques. Bull. Mar. Sci. 44: 1138-1151.

Dayton, P.K. 1971. Competition, disturbance, and community organization: The provision and subsequent utilization of space in a rocky intertidal community. Ecol. Monogr. 41: 351-389.

Dayton, P.K., G.A. Robilliard, R.T. Paine, and L.B. Dayton. 1974. Biological accommodation in the benthic community at McMurdo Sound, Antarctica. Ecol. Monogr. 44: 105-128.

Deehr, R.A., D.B. Barry, D.D. Chagaris and J.J. Luczkovich. 2007. Using SCUBA and snorkeling methods to obtain model parameters for an ecopath network model for Calabash Caye, Belize, Central America. In: Pollock, N.W. and J.M. Godfrey (eds.), Diving for Science 2007, Proceedings of Amer. Acad. of Underwater Sci. pp. 51-67.

DeMartini, E.E. and D. Roberts. 1982. An empirical test of biases in the rapid visual techniques for species-time censuses of reef fish assemblages. Mar. Biol. 70: 129-134.

Denley, E.J. and P.K. Dayton. 1985. Competition among macroalgae, pp. 512-530. In: Littler, M.M. and D.S. Littler (eds.). Handbook of Phycological Methods. Ecological Field Methods: Macroalgae. Cambridge University Press. 617 pp.

DeWreede, R.E. 1985. Destructive (harvest) sampling, pp. 147-160. In: Littler, M.M. and D.S. Littler (eds.). Handbook of Phycological Methods. Ecological Field Methods: Macroalgae. Cambridge University Press. 617 pp.

Di Santo, V., C. Pomory, and W.A. Bennett. 2009. Algal garden cultivation and guarding behavior of dusky damselfish on coral rubble and intact reef in Dry Tortugas National Park. In: Pollock, N.W. (ed.), Diving for Science 2009, Proceedings of Amer. Acad. of Underwater Sci. pp. 222-228.

Dowling, G. 1963. Diver's instrumented observation board. U.S. Navy Mine Defense Laboratory Report 210 (Arlington, VA, U.S. Navy Government Clearinghouse).

Downing, J.A. and M.R. Anderson. 1985. Estimating the standing biomass of aquatic macrophytes. Can. J. Aquat. Sci. 42: 1860-1869.

Drew, E.A. and B.P. Jupp. 1976. Some aspects of the growth of *Posidonia oceanica* in Malta. In: Drew, E.A., J.N. Lythgoe, and J.D. Woods (eds.). 1976. Underwater research. Academic Press, New York. pp. 357-367.

Drew, E.A., B.P. Jupp and W.A.A. Robertson. 1976. Photosynthesis and growth of *Laminaria hyperborea* in British Waters. In: Drew, E.A., J.N. Lythgoe, and J.D. Woods (eds.). Underwater research. Academic Press, New York. pp. 369-379.

Drew, E.A., J.N. Lythgoe, and J.D. Woods (eds.). 1976. Underwater research. Academic Press, New York. 430 pp.

Ducey, E.J. 2009. Assessing seasonal variation of epibenthic community structure on restored oyster reefs: Macroinvertebrate biodiversity and polychaete (*Nereis succinea*) biomass. In: Pollock, N.W. (ed.), Diving for Science 2009, Proceedings of Amer. Acad. of Underwater Sci. pp. 140-154.

Dupont, J.M. and C. Coy. 2008. Only the strong will survive: Red tides as community-structuring forces in the eastern Gulf of Mexico. In: Brueggeman, P. and N.W. Pollock, (eds.), Diving for Science 2008, Proceedings of Amer. Acad. of Underwater Sci. pp. 45-52.

Ebert, T.A. 1965. A technique for the individual marking of sea urchins. Ecology 46: 193-194.

Edwards, M. 1998. Proceedings of the 16th International Seaweeds Symposium.

Emery, L. and R. Wydoski. 1987. Marking and tagging of aquatic animals: An indexed bibliography. U.S. Fish and Wildlife Service Resource Publ. No. 165.

Engstrom-Heg, R. 1987. Persistence of rotenone in ponds. North Amer. J. Fish. Managm. 7: 162.

Ennis, G.P. 1972. A diver-operated plankton collector. J. Fish. Res. Bd. Can. 29: 341-343.

Fager, E., A Fleichsig, R. Ford, R. Clutter, and R. Ghelardi. 1966. Equipment for use in ecological studies using scuba. Limnol. and Oceanogr. 11(4): 503-509.

Feder, H.M. 1955. The use of vital stains in marking starfish. Science 85: 412.

Finnish IBP-PM Group. 1969. Quantitative sampling equipment for the littoral benthos. Int. Rev. Hydrobiol. 54: 185-93.

Forstner, H. and K. Rutzler. 1970. Measurements of the micro-climate in littoral marine habitats. Oceanogr. Mar. Biol. Ann. Rev. 8: 225-249.

Forsythe, J.W., A.R. Dinuzzo, and R.T. Hanlon. 1993. A method for qualitative assessment of bacterial flora on the skin of octopuses in nature. In: Heine, J.N. and N.L. Crane, (eds.), Diving for Science...1993, Proceedings of Amer. Acad. Underwater Sci. pp. 59-66.

Foster, M.S. 1975. Algal succession in a *Macrocystis pyrifera* forest. Mar. Biol. 32: 313-329.

Foster, M.S. 1976. A mini-slate for recording information underwater. Underwater Naturalist 9: 14-15.

Foster, M.S. Rhodoliths: between rocks and soft places. J. Phycol. 37: 659-667.

Foster, M.S., T.A. Dean, and L.E. Deysher. 1985. Subtidal Techniques, pp. 189-232. In: Littler, M.M. and D.S. Littler (eds.). Handbook of Phycological Methods. Ecological Field Methods: Macroalgae. Cambridge University Press. 617 pp.

Foster, M.S. and W.P. Sousa. 1985. Succession, pp. 269-290. In: Littler, M.M. and D.S. Littler (eds.). Handbook of Phycological Methods. Ecological Field Methods: Macroalgae. Cambridge University Press. 617 pp.

Gasca, R. and S.H.D. Haddock. 2004. Associations between gelatinous zooplankton and hyperiid amphipods (Crustacea: Peracarida) in the Gulf of California. Hydrobiologia. 530/531: 549-556.

Gerard, V.A. 1976. Some aspects of material dynamics and energy flow in a kelp forest in Monterey Bay, California. Ph.D. thesis, University of California, Santa Cruz. 173 pp.

Gilderhus, P.A. and L.L. Marking. 1987. Comparative efficacy of 16 anesthetic chemicals on rainbow trout. North Amer. J. Fish. Managm. 7: 288-292.

Gitschlag, G.R. 1986. A collapsible trap for underwater fish tagging. Bull. Mar. Sci. 39: 719-722.

Gitschlag, G.R. 1995. Assessment of fish mortality caused by explosive removal of offshore platforms in the Gulf of Mexico: A description of research diving and sampling techniques used on a NOAA project. In: Harper, D.E. (ed.), Diving for Science...1995, Proceedings of Amer. Acad. Underwater Sci. pp. 13-25.

Gochfeld, D.J. and G.S. Aeby. 2008. Antibacterial chemical defenses in Hawaiian corals provide possible protection from disease. Mar. Ecol. Prog. Ser. 362: 119-128.

Graham, K. and K.P. Sebens. 1996. The distribution of marine invertebrate larvae near vertical surfaces in the rocky subtidal zone. Ecology 77: 105-121.

Haaker, P.L., I. Taniguchi, and M. Artusio. 2005. Assessment of abalone stocks in southern California: The first stage of recovery. In: Godfrey, J.M. and S.E. Shumway (eds.), Diving for Science 2005, Proceedings of Amer. Acad. Underwater Sci. pp. 75-87.

Haddock, S.H.D. 2004. A golden age of gelata: Past and future research on planktonic ctenophores and cnidarians. Hydrobiologia. 530/531: 549-556.

Haddock, S.H.D. 2007. Comparative feeding behavior of planktonic ctenophores. Integr. Comp. Biol. 47(6): 847-853.

Haddock, S.H.D. and J.N. Heine. 2005. Scientific Blue-Water Diving Guidelines. Calif. Sea Grant Publ. No. T-057. Univ. of Calif., La Jolla, CA 92093. 49 pp.

Hall, S.J., D. Raffaelli, and W.R. Turrell. 1990. Predator-caging experiments in marine systems: A reexamination of their value. Am Nat. 77: 190-197.

Halusky, J.G. 1986. Training volunteer divers to research and document artificial reefs for their community. In: Mitchell, C.T. (ed.), Diving for Science...86, Proceedings of Amer. Acad. Underwater Sci. pp. 111-120.

Hampson, G.R., D.C. Rhoads, and D.W. Clark. 1989. Benthic mariculture and research rig developed for diver operations. In: Lang, M.A. and W.C. Jaap, (eds.), Diving for Science...1989, Proceedings of Amer. Acad. Underwater Sci. pp. 113-117.

Hanisak, M.D., S.M. Blair, and J.K. Reed. 1989. Use of photogrammetric techniques to monitor coral reef recovery following a major ship grounding. In: Lang, M.A. and W.C. Jaap, (eds.), Diving for Science...1989, Proceedings of Amer. Acad. Underwater Sci. pp. 119-135.

Hargreaves, N.B. 1992. An electronic hot-branding device for marking fish. Prog. Fish-cult. 54: 99-104.

Harlin, M.M. 1973. "Obligate" algal epiphyte: *Smithora naiadum* grown on a synthetic substrate. J. Phycol. 9: 230-232.

Harvey, E., D. Fletcher, M.R. Shortis, and G.A. Kendrick. 2004. A comparison of underwater visual distance estimates made by scuba divers and a stereo-video system: Implications for underwater visual census of reef fish abundance. Mar. and Freshw. Res. 55: 573-580.

Hebel, D. 1983. Concentration and flux of trace metals, carbon-nitrogen, and particulate matter in marine snow. Masters thesis, San Francisco State University, Moss Landing Marine Laboratories. 69 pp.

Heck, K.L. and T.A. Thoman. 1981. Experiments on predator-prey interactions in vegetated aquatic habitats. J. Exp. Mar. Biol. Ecol. 53: 125-134.

Heine, J.N. 1983. Seasonal productivity of two red algae in a central California kelp forest. J. Phycol. 19: 146-52.

Heine, J. N. 1989. Effects of ice scour on the structure of sublittoral marine algal assemblages of St. Lawrence and St. Matthew Islands, Alaska. Mar. Ecol. Progr. Ser. 52: 253-260.

Heine, J.N. (ed.). 1986. Blue water diving guidelines. Calif. Sea Grant Publ. No. T-CS-GCP-014. Univ. of Calif., A-032, La Jolla, CA 92093.

Herrlinger, T.J. 1983. The diet and predator-prey relationships of the sea star *Pycnopodia helianthoides* (Brandt) from a central California kelp forest. San Jose State University, Moss Landing Marine Laboratories, 57 pp.

Hidu, H. and J.E. Hanks. 1968. Vital staining of mollusk shells with alizarin sodium monosulfonate. Proc. Natn. Shellfish. Assn. 58: 37-41.

High, W.L. 1967. Scuba diving, a valuable tool for investigating the behavior of fish within the influence of fish gear. FAO Fish. Rep. (62)2, E2, 253-267.

High, W.L. 1998. Observations of a scientist/diver on fishing technology and fisheries biology. U.S. Dept. of Commerce, NOAA, Alaska Fisheries Science Center, National Marine Fisheries Service. Report 98-01, 48 pp.

Hill, J. and G.D. Grossman. 1987. Effects of subcutaneous marking on stream fishes. Copeia 1987(2): 492-495.

Hixon, M.A. 1991. Predation as a process structuring coral reef communities. In: Sale, P.F. (ed.), The ecology of fishes on coral reefs. Academic Press, San Diego, CA. pp. 475-508.

Hixon, M.A. and J.P. Beets. 1989. Shelter characteristics and Caribbean fish assemblages: Experiments with artificial reefs. Bull. Mar. Sci. 44: 666-680.

Hixon, M.A. and J.P. Beets. 1993. Predation, prey refuges, and the structure of coral-reef fish assemblages. Ecol. Monogr. 63(1): 77-101.

Hixon, M.A. and M.H. Carr. 1997. Synergistic predation causes density-dependent mortality in marine fish. Science 277: 946-949.

Hobson, E.S. and J.R. Chess. 1979. Zooplankters that emerge from the lagoon floor at night at Kure and Midway Atolls, Hawaii. Fish. Bull., U.S. 77: 275-280.

Hodges, V. and T.N. Frierson. 2008. Diving with captive sharks. In: Brueggeman, P. and N.W. Pollock (eds.), Diving for Science 2008, Proceedings of Amer. Acad. Underwater Sci. pp. 53-70.

Hoelzer, G.A. 1982. Movement patterns and the development of interactive behavior in juveniles of two territorial species of rockfish (Scorpaenidae: *Sebastes*). San Jose State University, Moss Landing Marine Laboratories, 53 pp.

Holme, N.A. and A.D. McIntyre (eds.). 1971. Methods for the study of marine benthos. Blackwell, Oxford. 334 pp.

Hsiao, S.I. and L.D. Druehl. 1973. Environmental control of gametogenisis in *Laminaria saccharina*. IV. *In situ* development of gametophytes and young sporophytes. J. Phycol. 9: 160-164.

Huber, B.T., J. Bijma, and H.J. Spero. 1996. Blue-water scuba collection of planktonic foraminifera. In: Lang, M.A. and C.C. Baldwin (eds.), Methods and Techniques of Underwater Science, Proceedings of Amer. Acad. Underwater Sci. pp. 127-132.

Hudson, J.H. 1972. Marking scallops with quick-setting cement. In: Proc. Nat. Shellfish Assoc., vol. 62.

Hulberg, L.W. and J.S. Oliver. 1980. Caging manipulations in marine soft-bottom communities: Importance of animal interactions or sedimentary habitat modifications. Can. J. Fish. Aquat. Sci. 37: 1130-1139.

Hulbert, S.H. 1984. Pseudoreplication and the design of ecological field experiments. Ecol. Monogr. 54: 187-211.

Jaap, W.C. and J. Wheaton. 1975. Observations on Florida coral reefs treated with fish-collecting chemicals. Fla. Mar. Res. Publ. 10: 1-18.

Jaap, W.C. 1986. A photogrametric apparatus for rapid areal benthic surveys. In: Mitchell, C.T. (ed.), Diving for Science...86, Proceedings of Amer. Acad. Underwater Sci. pp. 227-231.

Jakobsson, J. 1970. On fish tags and tagging. Mar. Biol. Ann. Rev. 8: 457-499.

James, P.W., S.C. Leon, A.V. Zale, and O.E. Maughan. 1987. Diver-operated electrofishing device. North Amer. J. Fish. Managm. 7: 597-598.

Jewett, S.C., R. Brewer, H. Chenelot, R. Clark, D. Dasher, S. Harper, and M. Hoberg. 2008. Scuba techniques for the Alaska Monitoring and Assessment Program (AKMAP) of the Aleutian Islands, Alaska. In: Brueggeman P. and N.W. Pollock (eds.). Diving for Science 2008, Proceedings of the American Academy of Underwater Sciences, pp. 71-89.

Johansen, H.W. and L.F. Austin. 1970. Growth rates in the articulated coralline *Calliarthron* (Rhodophyta). Can J. Bot. 48: 125-132.

John, D.M. and W. Pople. 1973. The fish grazing of rocky shore algae in the Gulf of Guinea. J. Exp. Mar. Biol. Ecol. 11: 81-90.

Johnston, C.A., I.S. Morrison, and K. MacLachlan. 1969. A photographic method for recording the underwater distribution of marine benthic organisms. J. Ecol. 57: 453-459.

Jones, R.S. and M.J. Thompson. 1978. Comparison of Florida reef fish assemblages using a rapid visual technique. Bull. Mar. Sci. 28: 159-172.

Kayes, R.J. 1974. The daily activity of *Octopus vulgaris* in a natural habitat. Mar. Behav. Physiol. 2: 337-343.

Kendall, J.J. and E.N. Powell. 1988. An *in situ* incubation procedure for examining the metabolic parameters of corals exposed to various stressing agents. In: M.A. Lang (ed.), Advances in Underwater Science...1988, Proceedings of Amer. Acad. Underwater Sci. pp. 77-88.

Kennelly, S.J. 1983. An experimental approach to the study of factors affecting colonization in a sublittoral kelp forest. J. Exp. Mar. Biol. Ecol. 68: 257-276.

Kennelly, S. and A. Underwood. 1984. Underwater microscopic sampling of a sublittoral kelp community. J. Exp. Mar. Biol. Ecol. 76: 67-78.

Kennelly, S. and A. Underwood. 1985. Sampling of small invertebrates on natural hard substrata in a sublittoral kelp forest. J. Exp. Mar. Biol. Ecol. 89: 55-68.

Kinne, O. and H.P. Burcheim (eds.). 1973. Man in the sea: *in situ* studies of life in oceans and coastal waters. Helgo. Meeres. 24: 1-535.

Kinsey, D.W. 1985. Open-flow systems, pp. 427-460. In: Littler, M.M. and D.S. Littler (eds.). Handbook of Phycological Methods. Ecological Field Methods: Macroalgae. Cambridge University Press. 617 pp.

Kline, T.C. and D. Scheel. 1996. Octopus research in Prince William Sound Alaska: The birthing of a scientific diving program and the role of the AAUS. In: Lang, M.A. and C.C. Baldwin (eds.), Methods and Techniques of Underwater Science, Proceedings of Amer. Acad. Underwater Sci. pp. 137-140.

Klos, E., J.H. Costello, and M.D. Ford. 1996. Measurement of the swimming behavior of three species of scyphomedusae in southern New England estuaries. In: Lang, M.A. and C.C. Baldwin (eds.), Methods and Techniques of Underwater Science, Proceedings of Amer. Acad. Underwater Sci. pp. 141-148.

Klos, E., J.H. Costello, S.P. Colin and W.M. Graham. 2005. Diving in two marine lakes in Croatia. In: Godfrey, J.M. and S.E. Shumway (eds.), Diving for Science 2005, Proceedings of Amer. Acad. Underwater Sci. pp. 211-216.

Kohler, K.E. and S.M. Gill. 2006. Coral Point Count with Excel extensions (CPCe): A Visual Basic program for the determination of coral and substrate coverage using random point count methodology. Computers and Geosci. 32 (9): 1259-1269.

Kooyman, G.L. 1966. Maximum diving capacities of the Weddell seal (*Leptonychotes weddelli*). Science 151: 1553.

Kooyman, G.L., C.M. Drabek, R. Elsner, and W.B. Campbell. 1971. Diving behavior of the Emperor penguin, *Aptenodytes forsteri*. The Auk 88: 775-795.

Langley, R.G. and J.W. Driscoll. 1989. Caudal peduncle tag used for identification of individual small fishes. Prog. Fish. Cult. 51: 55-57.

Langlois, T.J., E.S. Harvey, B. Fitzpatrick, J.J. Meeuwig, G. Shedrawi, and D.L. Watson. 2010. Cost-efficient sampling of fish assemblages: Comparison of baited video stations and diver video transects. Aquat. Biol. 9: 155-168.

Larson, R.J. and E.E. DeMartini. 1984. Abundance and vertical distribution of fishes in a cobble-bottom kelp forest off San Onofre, California. Fish. Bull. 82: 37-53.

Leaman, B.M. 1976. An inexpensive tag for short-term visual tracking studies. J. Fish. Res. Bd. Can. 33: 1628-1629.

Lees, D.C. 1968. Tagging subtidal echinoderms. Under. Nat. 5: 16-19.

Lessios, H.A. 1996. Methods for quantifying abundance of marine organisms. In: Lang, M.A. and C.C. Baldwin (eds.), Methods and Techniques of Underwater Science, Proceedings of Amer. Acad. Underwater Sci. pp. 149-157.

Liddell, W.D. and S.L. Ohlhorst. 1987. Patterns of reef community structure, north Jamaica. Bull. Mar. Sci. 40(2): 311-329.

Lincoln-Smith, M.P. 1988. Effects of observer swimming-speed on sample counts of temperate rocky reef fish assemblages. Mar. Ecol. Progr. Ser. 43: 223-231.

Lindholm, J., A. Knight, J. Btrantner, L. Kaufman, and S. Miller. 2009. Habitat-mediated signal reception by a passive acoustic receiver array as determined by scuba surveys. In: Pollock, N.W. (ed.), Diving for Science 2009, Proceedings of Amer. Acad. of Underwater Sci. pp. 75-85.

Lincoln-Smith, M. P. 1989. Improving multispecies rocky reef fish censuses by counting different groups of species using different procedures. Env. Biol. Fish. 26: 29-37.

Lindquist, D.G. and I.E. Clavijo. 1992. Nocturnal and crepuscular activity of reef fishes in Onslow Bay, N.C.: Scuba, video, and remotely operated vehicle observations. In: Cahoon, L.B. (ed.), Diving for Science...1992, Proceedings of Amer. Acad. Underwater Sci. pp. 99-107.

Littler, M.M., P.R. Taylor, and D.S. Littler. 1983. Algal resistance to herbivory on a Caribbean barrier reef. Coral Reefs 2: 111-118.

Littler, M.M. and K.E. Arnold. 1985. Electrodes and chemicals, pp. 349-375. In: Littler, M.M. and D.S. Littler (eds.). Handbook of Phycological Methods. Ecological Field Methods: Macroalgae. Cambridge University Press. 617 pp.

Littler, M.M. and D.S. Littler. 1985. Non-destructive sampling, pp. 161-176. In: Littler, M.M. and D.S. Littler (eds.). Handbook of Phycological Methods. Ecological Field Methods: Macroalgae. Cambridge University Press. 617 pp.

Lobel, P.S. 2008. Diver ecotourism and the behavior of reef sharks and rays - an overview. In: Brueggeman, P. and N.W. Pollock (eds.), Diving for Science 2008, Proceedings of Amer. Acad. of Underwater Sci. pp. 103-113.

Luckhurst, B.E. and K. Luckhurst. 1978. Analysis of the influence of substrate variables on coral reef fish communities. Mar. Biol. 49: 317-323.

Madin, L.P. 1974. Field observations on the feeding behavior of salps (Tunicata: Thaliacea). Mar. Biol. 25: 143-147.

Main, J. and G.I. Sangster. 1978. A new method for observing fishing gear using a towed net submersible. Progress in Underwater Science 3: 259-267.

Mann, C.G. 1989. Bioluminescence of gelatinous zooplankton in the Greenland and Barents Seas: Nightlights in the land of the midnight sun. In: M.A. Lang (ed.), Advances in Underwater Science...1988, Proceedings of Amer. Acad. Underwater Sci. pp. 229-239.

Mann, K.H. 1972. Ecological energetics of the seaweed zone in a marine bay on the Atlantic coast of Canada. 2. Productivity of the seaweeds. Mar. Biol. 14: 199-209.

Mann, K.H. and C. Mann. 1981. Problems of converting linear growth increments of kelps to estimates of biomass production. In: Levring, T. (ed.), Proc. Tenth Intern. Seaweed Symp. , pp. 699-704. DeGruyter, Berlin.

Markham, H.L. and N. Browne. 2007. Baseline survey protocol. In: Pollock, N.W. and J.M. Godfrey (eds.), Diving for Science 2007, Proceedings of Amer. Acad. of Underwater Sci. pp. 13-22.

Matsumoto, G.I. and G.R. Harbison. 1993. *In-situ* observations of foraging, feeding, and escape behavior in three orders of oceanic ctenophores: Lobata, Cestida, and Beroida. Mar. Biol. 117: 279-287.

Matthews, J. and J.D. Bell. 1979. A simple method for tagging fish underwater. Calif. Fish and Game 65: 113-117.

Mattison, J.E., J.D. Trent, A.L. Shanks, T.B. Akin, and J.S. Pearse. 1977. Movement and feeding activity of sea urchins (*Strongylocentrotus franciscanus*) adjacent to a kelp forest. Mar. Biol. 39: 25-30.

Max, M.D. and D.J. Nagel. 1997. Diving tablet and graphical relational database. In: Maney, Jr. E.J. and C.H. Ellis, Jr. (eds.), Diving for Science...1997, Proceedings of Amer. Acad. Underwater Sci. pp. 105-114.

McCormick, M.I. and J.H. Choat. 1987. Estimating total abundance of large temperate reef fish using visual strip-transects. Mar. Biol. 96: 469-478.

McErlean, A.J. and V.S. Kennedy. 1968. Comparison of some anesthetic properties of benzocaine and MS-222. Trans. Amer. Fish. Soc. 97: 496-498.

McFall, G.B., A.N. Shepard, C.L. Donaldson, and A.W. Hulbert. 1992. Development and application of a low-cost paired-laser measuring device. In: Cahoon, L.B. (ed.), Diving for Science...1992, Proceedings of Amer. Acad. Underwater Sci. pp. 109-113.

McMillan, S.M. 2008. Subtidal application of copper in the study of gastropod-algal interactions. In: Brueggeman, P. and N.W. Pollock (eds.), Diving for Science 2008, Proceedings of Amer. Acad. Underwater Sci. pp. 115-120.

Miles, E.L. and R.B. Whitlatch. 1997. "Priscilla": A portable *in situ* suction sampling device. In: Maney, Jr. E.J. and C.H. Ellis, Jr. (eds.), Diving for Science...1997, Proceedings of Amer. Acad. Underwater Sci. pp. 117-121.

Miller, R.J., D.C. Reed and M.A. Brzezinski. 2009. Community structure and productivity of subtidal turf and foliose algal assemblages. Mar. Ecol. Prog. Ser. 388: 1-11.

Miller, J.W. 1975. The NOAA Diving Manual: Diving for Science and Technology. NOAA, US Dept. of Commerce.

Mladenov, P.V. and I. Powell. 1986. A simple underwater magnifying device for the diving biologist. Bull. Mar. Sci. 38(3): 558-561.

Moring, J. 1987. Use of the anesthetic quinaldine for handling Pacific coast intertidal fishes. Trans. Amer. Fish. Soc. 99: 803-806.

Neushul, M. 1965. Diving observations of sub-tidal Antarctic marine vegetation. Bot. Mar. 8: 234-243.

Neushul, M., M.S. Foster, D.A Coon, J.W. Woessner, B.W.W. Harger. 1976. An *in situ* study of recruitment, growth and survival of subtidal marine algae: Techniques and preliminary results. J. Phycol. 12: 397-408.

Nicholson, N.L. 1970. Field studies on the giant kelp *Nereocystis*. J. Phycol. 6: 177-182.

North, W.J. 1971. Introduction and background: The biology of giant kelp beds (*Macrocystis*) in California. Nova Hedwigia 32: 1-97.

Olsson, M. and G. Newton. 1979. A simple, rapid method for marking individual sea urchins. Calif. Fish and Game. 65: 58-61.

Paquette, A.E., P.J. Auster, and M.D. Arendt. 2009. Approaches for analyzing behavioral interactions of fishes using time series video observations at an ocean observatory off the coast of Georgia. In: Pollock, N.W. (ed.), Diving for Science 2009, Proceedings of the Amer. Acad. of Underwater Sci., Sea Grant Publ. No. CTSG-10-09, pp. 206-215.

Perdokaris, C., C. Nathanailides, E. Gouva, U.U. Gabriel, K. Bitchava, F. Athanasopoulou, A. Paschou, and I. Paschos. 2010. Acta. Vet. Brno. 79: 481-490.

Pollock, E.G. 1969. Interzonal transplantation of embryos and mature plants of *Fucus*. In: Margalef, R. (ed.), Proc. Sixth International Seaweed Symp., pp. 345-356.

Pollock, N.W. and S.S. Bowser. 1995. Scuba collection of benthic foraminifera in Explorers Cove, Antarctica: An accessible model of the deep-ocean benthos. In: Harper, D.E. (ed.), Diving for Science...1995, Proceedings of Amer. Acad. Underwater Sci. pp. 63-74.

Postell, A., R.J. Rhodes, D. Swing, and B. Kennedy. 2008. In: Brueggeman, P. and N.W. Pollock (eds.), Diving for Science 2008, Proceedings of Amer. Acad. Underwater Sci. pp. 129-134.

Potts, T.A. and A.W. Hulbert. 1992. The use of fish aggregating devices (FADs) as an alternative to small-scale artificial reefs. In: Cahoon, L.B. (ed.), Diving for Science...1992, Proceedings of Amer. Acad. Underwater Sci. pp. 133-141.

Powers, S.P., D.E. Harper, and N.N. Rabalais. 1995. Underwater research methods for studying recruitment of benthic invertebrate larvae. In: Harper, D.E. (ed.), Diving for Science...1995, Proceedings of Amer. Acad. Underwater Sci. pp. 82-94.

Pringle, J.D. 1984. Efficiency estimates for various quadrat sizes used in benthic sampling. Can. J. Fish. Aquat. Sci. 41: 1485-1489.

Ragulin, A.G. 1969. Underwater observations on krill. Trudy VNIRO. 66: 231-234.

Randall, J.E. 1963. Methods of collecting small fishes. Underwater Naturalist 1: 6-11, 32-36.

Rasmussen, D. 1996. State mussel watch program 1993-95 data report. Calif. St. Water Res. Control Bd. Publ. 96-2WQ. 144 pp.

Raymond, H.L. 1974. Marking fishes and invertebrates. I. State of the art of fish branding. MFR paper 1067. Mar. Fish. Rev. 36:1-6; and II. Brand size and configuration in relation to long-term retention on steelhead trout and chinook salmon. MFR paper 1068. Mar. Fish. Rev. 36: 7-10.

Reed, D.C. and M.S. Foster. 1984. The effects of canopy shading on algal recruitment and growth in a giant kelp forest. Ecology 56: 937-948.

Reed, D.C., R.J. Lewis and M. Anghera. 1994. Effects of an open coast oil outfall on patterns of giant kelp (*Macrocystis pyrifera*) recruitment. Mar. Biol. 120: 25-31.

Reed, K.C., E.M. Muller, and R. van Woeslik. 2010. Coral immunology and resistance to disease. Dis. Aquat. Org. 90: 85-92.

Reed, S.A. 1980. Sampling and transecting techniques on tropical reef substrates. In: Environmental survey techniques for coastal water assessment conference proceedings. University of Hawaii Sea Grant College Program, Cooperative Report UNIHI-SEA-GRANT-CR-80-01. pp. 71-89

Richardson, L.L. 1992. Red band disease: A new cyanobacterial infestation of corals. In: Cahoon, L.B. (ed.), Diving for Science...1992, Proceedings of Amer. Acad. Underwater Sci. pp. 153-160.

Ridgway, S.H. 1972. Mammals of the sea: Biology and medicine. Charles C. Thomas, Publ. 812 pp.

Runnels, R. J. 1985. Diver sampling of macroinvertebrates on northwestern Gulf of Mexico hard bottom areas. In: Mitchell, C.T. (ed.), Diving for Science...85, Proceedings of Joint International Scientific Diving Symposium, Amer. Acad. Underwater Sci. pp. 115-122.

Russell, B.C., F.H. Talbot, G.V.R. Anderson, and B. Goldman. 1978. Collection and sampling of reef fishes, pp. 329-345. In: Stoddart, D.R. and R.E. Johannes (eds.). Coral Reefs: Research Methods. UNESCO, Paris.

Rutzler, K. 1978. Sponges in coral reefs. In: Stoddart, D.R. and R.E. Johannes (eds.) Coral Reefs: Research Methods, Monographs on Oceanographic Methodology 5. Paris, UNESCO. pp. 299-313.

Rutzler, K. 1996. Sponge diving - professional but not for profit. In: Lang, M.A. and C.C. Baldwin (eds.), Methods and Techniques of Underwater Science, Proceedings of Amer. Acad. Underwater Sci. pp. 183-204.

Rutzler, K., J.D. Ferraris and R.J. Larson. 1980. A new plankton sampler for coral reefs. P.S.Z.N.I. I: Mar. Ecol. 1: 65-71.

Sale, P.F. 1971. Extremely limited home range in a coral reef fish, *Dascyllus aruanus* (Pisces: Pomacentridae). Copeia. 1971: 324-327.

Sale, P.G. and W.A. Douglas. 1981. Precision and accuracy of visual census techniques for fish assemblages on coral reef patch reefs. Env. Biol. Fish. 6: 333-340.

Sale, P. G. and B. J. Sharp. 1983. Correction for bias in visual transect censuses of coral reef fishes. Coral Reefs 2: 37-42.

Sanders, G.S. and F.E. Wendell. 1991. Closed-circuit oxygen apparatus: Minimizing risks for improved efficiency. In: Krock, H.J. and D.E. Harper (eds.), International Pacifica Scientific Diving...1991, Proceedings of Amer. Acad. Underwater Sci. pp. 87-101.

Sanderson, S.L. and S.C. Solonsky. 1986. Comparison of a rapid visual and a strip transect technique for censusing reef fish assemblages. Bull. Mar. Sci. 39: 119-129.

Savy, S. 1987. Activity pattern of the sea star *Marthasterias glacialis* in Port-Cros Bay (France, Mediterranean coast). Mar. Ecol. 8: 97-106.

Scarborough-Bull, A. and J.J. Kendall. 1992. Preliminary investigation: Platform removal and associated biota. In: Cahoon, L.B. (ed.), Diving for Science...1992, Proceedings of Amer. Acad. Underwater Sci. pp. 31-37.

Scarborough-Bull, A. 1993. Study planning, testing, and reality: Platform removal and associated biota. In: Heine, J.N. and N.L. Crane, (eds.), Diving for Science...1993, Proceedings of Amer. Acad. Underwater Sci. pp. 117-125.

Schick, D.F., L. Mayer, J. Riley, and G. Gust. 1988. Use of a flume for sampling the sediment-water interface. In: M.A. Lang (ed.), Advances in Underwater Science...1988, Proceedings of Amer. Acad. Underwater Sci. pp. 133-141.

Schmidt, G.H. and G.F. Warner. 1984. Effects of caging on the development of a sessile epifaunal community. Mar. Ecol. Progr. Ser. 15: 251-263.

Schmitt, P.J. 1984. Marking growth increments in otoliths of larval and juvenile fish by immersion in tetracycline to examine the rate of increment formation. Nat. Mar. Fish. Serv. Fish Bull. 82: 237-242.

Schmitt, E.F. and K.M. Sullivan. 1996. Analysis of a volunteer method for collecting fish presence and abundance data in the Florida Keys. Bull Mar Sci. 59(2): 404-416.

Schroeder, W. 1974. Collecting and handling zooplankton and epibenthic organisms underwater. Mar. Tech. Soc. J. 8(5): 40-43.

Sebens, K.P. 1976. Individual marking of soft-bodied intertidal invertebrates *in situ*; a vital stain technique applied to the sea anemone *Anthopleura xanthogrammica*. J. Fish. Res. Bd. Can. 33: 1407-1410.

Sebens, K.P., E.J. Maney, and J.H. Witting. 1992. A portable, diver-operated plankton sampler for near substratum use. In: Cahoon, L.B. (ed.), Diving for Science...1992, Proceedings of Amer. Acad. Underwater Sci. pp. 167-172.

Sebens, K.P., E.J. Maney, Jr., and A. Gordon. 1997. Long term research in the rocky subtidal zone (Massachusetts 1977-1997). In: Maney, Jr. E.J. and C.H. Ellis, Jr. (eds.), Diving for Science...1997, Proceedings of Amer. Acad. Underwater Sci. pp. 141-159.

Shearer, T.L. 2009. Population analysis of an introduced coral species, *Tubastrea coccinea*, in Florida. In: Pollock, N.W. (ed.), Diving for Science 2009, Proceedings of Amer. Acad. of Underwater Sci. pp. 229-236.

Shepherd, S. 1985. Power and efficiency of a research diver, with a description of a rapid underwater measuring gauge: Their use in measuring recruitment and density of an abalone population. In: Mitchell, C.T. (ed.), Diving for Science...85, Proceedings of Joint International Scientific Diving Symposium, Amer. Acad. Underwater Sci. pp. 263-272.

Sigl, W., V. VonRad, H.J. Oeltzschner, K. Braune, and F. Fabricius. 1969. Diving sled: A tool to increase the efficiency of underwater mapping by scuba divers. Mar. Geol. 7: 357-363.

Smith, C.L. 1973. Small rotenone stations: A tool for studying coral reef fish communities. Amer. Mus. Novit. 2512: 1-21.

Smith, D., T. Kornfield and B. Goodwin. 1981. Remotely operated trap for capture of territorial fishes. Prog. Fish Cult. 43: 208-209.

Somers, L.H. 1990. The University of Michigan Diving Manual, Vol. II: Underwater Research Methods. Ann Arbor, MI. 182 pp.

Spanier, E., M. Tom, S. Pisanty, S. Breitstein, Y. Tur-Caspa, and G. Almog. 1985. Cold stress and the scientific diver. In: Mitchell, C.T. (ed.), Diving for Science...85, Proceedings of Joint International Scientific Diving Symposium, Amer. Acad. Underwater Sci. pp. 123-135.

Starr, R.M., J.N. Heine, J.M. Felton, and G.M. Cailliet. 2002. Movements of bocaccio (*Sebastes paucipinus*) and greenspotted (*S. chlorostictus*) rockfishes in a Monterey submarine canyon: Implications for the design of marine reserves. Fish. Bull. 100: 324-37.

Stasko, A.B. and D.G. Pincock. 1977. Review of underwater telemetry, with emphasis on ultrasonic techniques. J. Fish. Res. Bd. Can. 34: 1261-1285.

Stretch, J.J. 1985. Quantitative sampling of demersal zooplankton: Re-entry and airlift dredge sample comparisons. J. Exp. Mar. Biol. Ecol. 91: 125-136.

Swenson, W.A., W.P. Gobin, and T.D. Simonson. 1988. Calibrated mask-bar for underwater measurement of fish. North Amer. J. Fish. Manag. 8: 382-385.

Tanner, C., M.W. Hawkes, and P.A. Lebednik. 1977. A hand-operated suction sampler for the collection of subtidal organisms. J. Fish. Res. Bd. Can. 34: 1031-1034.

Taylor, R.B., R.A. Blackburn, and J.H. Evans. 1995. A portable battery-powered suction device for the quantitative sampling of small benthic invertebrates. J. Exp. Mar. Biol. Ecol. 194: 1-7.

Tengberg, A., F. de Bovee, P. Hall, W. Berelson, D. Chadwick, G. Ciceri, P. Crassous, A. Devol, S. Emerson, J. Gage, R. Glud, F. Graziottin, J. Gundersen, D. Hammond, W. Helder, L. Hinga, O. Holby, R. Jahnke, A. Khripounoff, S. Lieberman, V. Nuppenau, O. Pfannkuche, C. Reimers, G. Rowe, A. Sahami, F. Sayles, M. Schurter, D. Smallman, B. Wehrli, and P. de Wilde. 1995. Benthic chamber and profiling landers in oceanography -- a review of design, technical solutions and functioning. Prog. Oceanogr. 35: 253-292.

Thayer, C.W. 1985. Quick-release cages and repetitive censusing of sessile epifauna. J. Exp. Mar. Biol. Ecol. 94: 251-257.

Thompson, K.W., L.A. Knight, and N.C. Parker. 1986. Color-coded fluorescent plastic chips for marking small fishes. Copeia (2): 544-545.

Thresher, R.E. and A.M. Gronell. 1978. Subcutaneous tagging of small reef fishes. Copeia (2): 352-353.

Thresher, R.E. and J.S. Gunn. 1986. Comparative analysis of visual census techniques for highly mobile, reef-associated piscivores (Carangidae). Env. Biol. Fishes 17: 93-116.

Towle, D.W. and J.S. Pearse. 1973. Production of the giant kelp, *Macrocystis*, estimated by *in situ* incorporation of ^{14}C in polyethylene bags. Limnol. Oceanogr. 18: 155-159.

Trevelyan, G.A. and E.S. Chang. 1987. Light-induced shell pigmentation in post-larval *Mytilus edulis* and its use as a biological tag. Mar. Ecol. Progr. Ser. 39: 137-144.

Tronzo, C.R., L.B. Cahoon, and D. B. Freeman. 1986. The quantitative sampling of demersal zooplankton in Onslow Bay, North Carolina. In: Mitchell, C.T. (ed.), Diving for Science...86, Proceedings of Amer. Acad. Underwater Sci. pp. 163-170.

Tsuda, R.T., and I.A. Abbott. 1985. Collection, handling, preservation, and logistics. pp. 67-86. In: Littler, M.M. and D.S. Littler (eds.). Handbook of Phycological Methods. Ecological Field Methods: Macroalgae. Cambridge University Press. 617 pp.

Tunberg, B. 1983. A simple, diver-operated suction sampler for quantitative sampling in shallow, sandy-bottoms. Ophelia 22: 185-188.

Tunnell, J.W. and T.J. Nelson. 1989. A high density - low diversity octocoral community in the southwestern Gulf of Mexico. In: M.A. Lang (ed.), Advances in Underwater Science ...1988, Proceedings of Amer. Acad. Underwater Sci. pp. 325-335.

Tusting, R.F. and D.L. Davis. 1993. Improved methods for visual and photographic benthic surveys. In: Heine, J.N. and N.L. Crane, (eds.), Diving for Science...1993, Proceedings of Amer. Acad. Underwater Sci. pp. 157-172.

Tutschulte, T.C. 1968. Monitoring the nocturnal movement of abalones. Underwater Nat., Bull. Am. Littoral Soc. 4(3): 12-15.

Vadas, R.L. 1985. Herbivory. pp. 531-572. In: Littler, M.M. and D.S. Littler (eds.). Handbook of Phycological Methods. Ecological Field Methods: Macroalgae. Cambridge University Press. 617 pp.

Valz, J.H. and P.A. Dinnel. 2007. Bottomfish variability in the proposed marine reserves of Skagit County, Washington. In: Pollock, N.W. and J.M. Godfrey (eds.), Diving for Science 2007, Proceedings of Amer. Acad. of Underwater Sci. pp. 1-12.

Virnstein, R.W. 1978. Predator caging experiments in soft sediments: Caution advised. In: M.L. Wiley (ed.), Estuarine Interactions. Academic Press, New York. pp. 261-273.

von Brandt, A. 1964. Fish catching methods of the world. Fishing News (books) Ltd., London, 191 pp.

Waaland, S.D. and J.R. Waaland. 1975. Analysis of cell elongation in red algae by fluorescent labeling. Planta 126: 127-138.

Watson, D.L., E.S. Harvey, B.M. Fitzpatrick, T.J. Langlois and G. Shedrawi. 2010. Assessing reef fish assemblage structure: How do different stereo-video techniques compare? Mar. Biol. 157: 1237-1250.

Wedemeyer, G. 1970. Stress of anesthesia with MS-222 and benzocaine in rainbow trout (*Salmo gairdneri*). J. Fish. Res. Bd. Can. 27: 909-914.

Weispfenning, A.J., P.A. Dinnel, N.T. Schwarck, and G. McKeen. 2005. Baseline scuba assessments of habitat and fishery resources in eight candidate marine reserve sites in Skagit County, Washington. In: Godfrey, J.M. and S.E. Shumway (eds.), Diving for Science 2005, Proceedings of Amer. Acad. Underwater Sci. pp. 129-142.

Wickham, D.A. and J.W. Watson, Jr. 1976. Scuba diving methods for fishing system evaluation. Mar. Fish. Rev. 1192: 15-23.

Wilson, T.C. 1981. An underwater fish tagging method. Calif. Dept. Fish and Game, 67: 47-50.

Winter, J.D. 1983. Underwater biotelemetry, pp. 371-393. In: Nielsen, L.A. and D.L. Johnson (eds.), Fisheries techniques. Amer. Fish. Soc., Bethesda, MD. 468 pp.

Witman, J.D. 1985. Refuges, biological disturbance, and rocky subtidal community structure in New England. Ecol. Monogr. 55: 421-445.

Witman, J.D. 1987. Subtidal coexistence: storms, grazing, mutualism, and the zonation of kelps and mussels. Ecol. Monogr. 57: 167-187.

Wolcott, T.G. and A.H. Hines. 1996. Advances in ultrasonic biotelemetry for animal movement and behavior: The blue crab case study. In: Lang, M.A. and C.C. Baldwin (eds.), Methods and Techniques of Underwater Science, Proceedings of Amer. Acad. Underwater Sci. pp. 229-236.

Woodin, S.A. 1978. Refuges, disturbance, and community structure: A marine soft bottom example. Ecology 59: 274-284.

Woods, J.D. and N. Lythgoe (eds.). 1971. Underwater science. Oxford University Press, London. 330 pp.

Workman, I.K. 1986. Trawl diving: A method used by fishing gear technologists to evaluate trawling systems. In: Mitchell, C.T. (ed.), Diving for Science...86, Proceedings of Amer. Acad. Underwater Sci. pp. 233-238.

Yoklavich, M.M., G.M. Cailliet, and G. Moreno. 1993. Rocks and Fishes: Submersible observations in a submarine canyon. In: Heine, J.N. and N.L. Crane, (eds.), Diving for Science...1993, Proceedings of Amer. Acad. Underwater Sci. pp. 173-181.

Youngbluth, J.J. 1982. Sampling demersal zooplankton: A comparison of field collections using three different emergence traps. J. Exp. Mar. Biol. Ecol. 61: 111-124.

Yund, P.O., S.D. Gaines, and M.D. Bertness. 1991. Cylindrical tube traps for larval sampling. Limnol. Oceanogr. 36: 1167-1177.

Zardus, J.D. 1997. Methods of a subtidal experiment involving infaunal density manipulations and sampling with replacement. In: Maney, Jr. E.J. and C.H. Ellis, Jr. (eds.), Diving for Science...1997, Proceedings of Amer. Acad. Underwater Sci. pp. 161-169.

Zieman, J.C. 1974. Methods for studies on the marine alga *Iridaea* and *Gigartina*. J. Exp. Mar. Biol. Ecol. 11: 71-80.

chapter 8

UNDERWATER PHOTOGRAPHY AND VIDEOGRAPHY FOR THE SCIENTIST

PHOTOGRAPHIC OR VIDEOGRAPHIC DOCUMENTATION of underwater research is an important component of many scientific projects (Foster, 1975; Done, 1981; de Strobel, 1985; Hanisak et al., 1989; Kendall and Bright, 1989; Maney et al., 1990; Harvey and Shortis, 1996; Goldberg and Foster 2002; Barans et al., 2005; Barber and Auster, 2005; Jewett et al., 2008; Paquette et al., 2009; Vroom et al., 2010). Photodocumentation can aid the scientist in many ways, such as recording behavioral observations, for photogrametric determination of species distribution and abundance, as a tool for taxonomic determinations, for evaluating instrument design and performance, and as a record of methods used in underwater experiments. Many scientists give presentations of their work at meetings or workshops, and underwater photographs of study sites, organisms, and methods are often very useful. There is a wide variety of equipment available to aid the scientist in achieving his or her objectives, ranging from close-up and wide-angle photography to time-lapse videography.

Photography and videography have undergone a tremendous change in technology and format in the past 15 years, from the predominance of 35 mm film SLR cameras and movie film to digital technology. The original and perhaps the most popular underwater camera for many years was the Nikonos, manufactured by Nikon, Inc. The first Nikonos appeared on the market in the 1960s and various models and improvements have been made over the years. On older models of Nikonos cameras (Nikonos I, II, and III), it was necessary to remove the lens before opening the camera body. These models also only have manual exposure settings. The newer models (Nikonos IV and V) have hinged camera backs, similar to land cameras, and offer automatic exposure operation.

The last model is the Nikonos V 35 mm aperture-priority automatic camera, which was available until 2001. The Nikonos V is a direct-vision viewfinder type camera, which means that you do not focus through the lens, but estimate the focal distance and set it manually. Exposure can be manual or automatic, using the light intensity feedback metering system.

The lenses manufactured by Nikon for the Nikonos V and older models include an 80 mm and 35 mm, which can be used on land and underwater, and 28 mm, 20 mm, and 15 mm lenses, which are exclusively for underwater use. Optical viewfinders are also available to allow for more precise composition. For close-up photography, extension tubes or a close-up kit can be used.

After the Nikonos V was discontinued, Nikon produced the Nikonos RS, a completely amphibious SLR camera. It was an excellent camera, with many features, but was very expensive, and prone to flooding. It was discontinued in the late 1990s.

Many of the limitations of traditional SLR film cameras, such as limited number of exposures per dive (typically 36), uncertainty of results until film is processed (could be weeks later), and difficulty in publication, transmittal, or display options, have been alleviated with digital technology. Instead of capturing images on film, digital cameras use electronic sensors, which store the image as a digital file. The images can be viewed instantly on the LCD screen on the camera. When downloaded to a personal computer, these files can then be processed and manipulated for color correction, sizing, or composition. Another important feature for scientific applications is the ability to embed metadata, such as time, date, camera model, and exposure, within the image file.

FIGURE 8.1:
Nikonos V underwater camera with strobe. (photo courtesy of J. Heine)

PHOTOGRAPHIC EQUIPMENT

Digital Cameras and Lenses

DIGITAL CAMERAS ARE RELATIVELY INEXPENSIVE, and can hold hundreds of images in memory. There are two broad categories of cameras that can be used underwater: waterproof cameras that are designed to be taken underwater, and traditional land cameras that are used in a waterproof housing. The simplest and least expensive underwater cameras are the "point and shoot" digital cameras that are rated to depths of about 10-30 ft. (3 to 10 m). The Fujifilm XP10 offers 12.2 megapixels of resolution, 720p HD video, a 5X optical zoom lens, and a 2.7-inch LCD. The XP10, which is called the XP11 in some parts of the world, measures only 0.9 inches in thickness. It also offers some "tough" features, including the ability to function in up to 10 feet of water, tolerate a 3-foot drop, and 14°F temperatures. The Olympus Stylus Tough-8010 has 14 megapixels of resolution, 720p HD video, a 5X optical zoom lens, and is waterproof to 10 m (33 ft). Panasonic offers a similar model called the TS10, Canon has the PowerShot D10, and Pentax offers a model called the Optio W90. SeaLife has a self-contained underwater digital camera that is depth rated to 200 ft (60 m). The DC1200 Elite has 12 megapixels resolution, video mode, 25X total zoom, and will accept an external flash.

Putting a digital camera in a waterproof housing allows for more options and increased depth ratings. There are a variety of waterproof housings available for almost every digital camera. Point and shoot digital cameras for use underwater should have wide-angle lens capabilities and manual exposure options. Popular models include the Olympus FE-360, Canon SD-940, S90 and G11, and Sea&Sea 1200HD. Sea&Sea also has a model called the DX-GE5 with 12 megapixels resolution, a 4X optical zoom lens, and is waterproof to a depth of 55 m (180 ft). The DX-2G offers an improved macro mode for close-up photography.

Beyond the point and shoot digital cameras are digital SLR cameras, which have additional features for advanced photography. Like the older film-version SLRs, these camera bodies can be fitted with a variety of lenses to suit the situation being photographed. They also have higher resolution, full HD movie mode, faster autofocus, and often larger LCD screens.

FIGURE 8.2: Digital camera in waterproof housing. (photo courtesy of D. Kesling)

Strobe Lights

FOR MOST WATER CONDITIONS AND APPLICATIONS, artificial light must be supplied by using a strobe. There are many makes and models of underwater strobe lights that can be used with digital cameras. Many strobes accommodate the through-the-lens (TTL) automatic exposure system, and allow for manual settings as well. There are a number of other strobes, which use either alkaline, or rechargeable nickel-cadmium (ni-cad) batteries.

The power or intensity of the strobe is given by its guide number. The higher the guide number, the more powerful the strobe. This guide number is also used to estimate manual exposure settings. For example, a strobe may have a guide number of 30 for a certain equivalent film speed. If you divide the subject to camera distance into the guide number, this gives you the approximate f stop (aperture) to set on the camera. For example, if the distance is three feet, then 30/3 = 10. The closest f stop to 10 is f11, so that would be the preferred aperture. Some strobes have multiple power settings, which can be used to aid in exposure and composition.

Some underwater photographers prefer to use two strobes to avoid the shading that can occur with one strobe. This is especially important in quantitative photography (see below). Both strobes can be connected to the camera, or one strobe can be a "slave," which means that it is triggered to fire when the primary strobe flashes.

The Sea and Sea YS-250 is one of the brightest strobes available, with a guide number of 32 and a fast recycle time. The Ikelite DS-161 strobe recycles very quickly and can also be used in a movie mode for video. The Inon S-2000 is a compact strobe that is powered by 4 AA batteries.

Underwater Camera Housings

MANY MODELS OF DIGITAL CAMERAS CAN BE USED underwater with a watertight housing (such as Aquatica, Amphibico, Fantasea, Fisheye, Ikelite, Nauticam, Sea&Sea, Subal). Some of the advantages of this type of system include improved composition of images, manual focus option, a variety of exposure metering choices, zooming from wide angle to telephoto, underwater playback, and one-button movie mode. Many SLRs are extremely sophisticated, and offer many more functions than most non-housed underwater cameras, but also cost considerably more.

Relatively inexpensive flexible plastic housings are manufactured by Ewa-Marine, and are tested to a depth of 30 m (100 ft). Full-featured models of housings are constructed of polycarbonate or aluminum and can accommodate TTL (automatic flash adjustment) via a fiber optic link. External strobes are triggered by the camera's built-in flash. Dome ports are available for wide-angle photography.

For a listing of cameras, housings, strobes, and other accessories, see the manufacturers at the end of this chapter.

FIGURE 8.3:
Underwater housings can utilize large dome ports for wide-angle photography.
(photo courtesy of B. Carlson)

Quantitative Photography Techniques

THE ADVANTAGES OF QUANTITATIVE PHOTOGRAMMETRIC techniques are well-documented in Littler and Littler (1985). This non-destructive sampling has minimal impact on a community, is relatively rapid, provides a permanent record of the community at discreet times, and shows a high level of precision and repeatability. However, small organisms are not easily quantified, and dense, multi-layered algal canopies can be difficult to quantify.

Dayton et al. (1974) describe the extensive use of photo-transects to evaluate the abundance, size frequency, and space utilization of sponges in Antarctica. This method was particularly useful at greater depths where time was limited. They took photographs at 1 m intervals along a 20 m line. The photographer was approximately 2-3 m above the substratum, which produced a photograph of 6-9 m^2 in area. To eliminate parallax problems, only the center 1 m^2 was analyzed. Percent cover of each species was determined by using a planimeter or random dots.

Blair and Flynn (1989) describe a continuous belt transect of slightly overlapping photographs that are 70 cm x 70 cm, yielding a 14 m^2 sampling area. Smaller, cryptic organisms are sampled using smaller quadrats of 40 cm x 40 cm.

A number of researchers describe using a "quadrapod," which is a camera mounted on a frame that is attached to a quadrat (Reed, 1980; Hanisak et al., 1989; Coyer and Witman, 1990; Jaap et al., 1990; Sebens et al., 1997; Coyer et al., 1999). A digital camera with a wide-angle lens and strobe(s) are mounted to a four-legged frame which has the desired

FIGURE 8.4:
Quadrapod framer with camera, strobes, gauges, and optional divided quadrat.
(photo courtesy of G. Eyal)

quadrat built into the bottom end. The quadrapod is held at right angles to the substrate and a photograph is taken. If layering of organisms is present, a series of photographs of the same quadrat can be taken, with the canopy moved to one side each time. For details on scoring the photoquadrat images for percent cover, see Littler and Littler (1985), Hanisak et al. (1989), and the description in Training Exercise #11.

Photographs taken with a digital camera allow for more precise evaluation of density or percent cover using image analysis software. Sizes of organisms can also be derived directly from photographs, provided that a scale or ruler is used in the quadrat when the photograph is taken. Quantitative photography is also useful for demographic studies, where the number of newly settled and surviving organisms can be tracked over time (Goldberg and Foster, 2002). Small organisms will not be easily detectable when using wide angle lenses for quantitative photography.

TRAINING EXERCISE #11
Photogrametric Transect for Benthic Population Estimations

Equipment needed: transect tape, underwater camera with attached framer or quadrats
Objective: To quantitatively evaluate the percent cover and/or density of macroflora and fauna

Ideally, there will be one camera, transect, and quadrat per diving buddy group, but if not available, the group can share the equipment. Prior to the dive, the staff should assign to each group random compass bearings for laying out the transect tape, and random distances along the tape to sample photographically. The divers will lay out the tape, and proceed to photograph each assigned quadrat. If the camera has a framer attached, the framer should be placed as flush to the substrate as possible. If using a quadrat, the camera must be held a fixed distance away from the quadrat. This can be repeatedly accomplished by using a meter stick or weighted piece of line. To avoid problems with improper distance, holding the camera at the wrong angle to the substrate, or strobe(s) aimed incorrectly, a quadrapod can be utilized (see Coyer and Witman, 1990; Coyer et al., 1999 for details). Each diver should practice taking photographs.

In the laboratory, to estimate percent cover and density, the images can be projected onto a sheet of paper with 100 random dots. Each dot is "scored" as a hit on a given species. The total number of hits can then be used to generate a percent cover for each species. Large, readily identifiable organisms can also be counted in each image, and knowing the area of the quadrat or framer, species densities can be obtained (see Littler and Littler, 1985; and Coyer and Witman, 1990, for details). Image analysis software can also be utilized.

Stereographic Photography

EARLY ACCOUNTS OF STEREOPHOTOGRAPHIC METHODS for quantitative studies on sublittoral rocky substrates are given by Lundalv (1971, 1976). Sampling is accomplished by hanging a photographic frame on dowels placed into holes drilled in the rock wall. A camera is hung on the frame, and two exposures are taken 20 cm apart, giving a stereoscopic pair

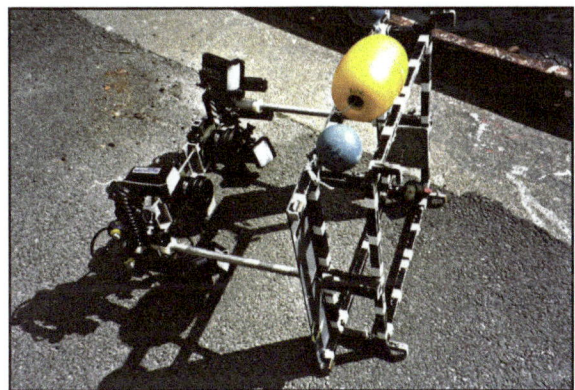

FIGURE 8.5:

Stereophotographic camera apparatus with flashes, depth gauge, and inclinometer.
(photo courtesy of M. Walday)

FIGURE 8.6:

3D HD camera system used by the National Park Service.
(photo courtesy of S. Pershern)

of photographs. The stereophotographs were analyzed with a Wild microstereo comparator, giving species composition, density, percent cover, size of organisms, and spatial patterns. A more current method of stereophotography and photogrammetry is available, which produces matched pairs of photographs that are used to determine the dimension and shape of objects (Jaap, 1986; Shortis et al., 2009). Two underwater cameras with strobes are mounted on a frame that is 1.2 m (4 ft.) above the substrate, which creates photographs of 1m by 0.75m (3 ft. by 2.3 ft.) of the sea floor (Done, 1981).

In the stereophotography design pictured in Figure 8.5, two mechanically synchronized Nikonos V cameras with 15 mm lenses are mounted on a frame that is attached to a 0.25 m^2 quadrat. Two stereomicroscopes are used to score the slides. A quantitative analysis can be done using a point-contact method. Information on size and growth over time can also be obtained using this apparatus.

High-definition three-dimensional (3D) cameras are used by the National Park Service Submerged Resources Center and Woods Hole Oceanographic Institution. The footage was used to produce a film on the *USS Arizona* Memorial, as most visitors do not see much of the shipwreck which is underwater. It is also being used in other National Parks around the country.

VIDEOGRAPHIC EQUIPMENT

Digital Video Cameras

EARLY FILM AND VIDEO cameras used roll film and video tape and were bulky and rather expensive. Current digital technology provides scientists with a relatively inexpensive and readily available means to record events underwater. High resolution video cameras in waterproof housings can be hand-held by divers, mounted on a base or onto the substrate, or deployed on ROV's for use in areas that are not accessible to divers. Housings can be constructed from polycarbonate, plastic, PVC, or aluminum (e.g. Amphibico, Ikelite, Light and Motion, Nimar, Sea and Sea, Sony, EWA Marine).

FIGURE 8.7:
Early scientific divers such as Conrad Limbaugh (shown here in the early 1950's) used a motion picture camera in a waterproof housing to film cleaning behavior of tropical reef animals.
(photo courtesy of W. North)

Video photography has advantages such as long recording times, time-lapse options, continuous recording, instant/real time viewing of the captured images, image stabilization, and a permanent record of transects. Most housings and cameras allow for use of controls such as on/off, record, focus, zoom, white balance, and exposure control. Additionally, some have external microphones, moisture sensors, and moveable red filters. A variety of LCD screens are available in various sizes, but the larger they are, the faster they use up battery life.

FIGURE 8.8:
Modern video cameras can be used in PVC housings.
(photo courtesy of J. Heine)

Digital videography has many of the same features as found in digital still cameras. First generation digital camcorders use small cassette tapes and can record high definition video on MiniDV tape. Modern camcorders are classified as standard definition, with a resolution of 720x480 pixels, or high definition, with a resolution of 1280x720 pixels or higher. They save video to internal flash memory, a hard drive, or SD memory cards. Many have the ability to film in very low light.

Most camcorders have USB 2.0 ports for transferring video files to a personal com-

FIGURE 8.9:
Diver using HD video camera to document Elkhorn coral.
(photo courtesy of B. Seymour)

puter. Many have HDMI ports as well, to allow direct connection to HDTV for playback. The most common format for video files is MPEG-4. Ultracompact waterproof camcorders are available that feature full high definition recording capability down to a depth of 3 m (10 ft.) (Kodak Playsport Zx3). They lack many of the features that are necessary for serious scientific work.

Light and Motion Industries makes a very compact video housing called the Stingray2, which supports most HD Canon and Sony camcorders. It features finger tip handle controls and pre-programmed one touch white balance controls. The Bluefin model allows for a variety of all glass lenses to be used, including a Fisheye 145 lens.

Video lights are useful under low light conditions, and to add color in blue and deeper waters. Depending upon the wattage and battery pack, burn times of greater than one hour can be achieved. Nickel cadmium (Nicad) and longer-lasting nickel metal hydride (NiMH) rechargeable batteries are available for most video lighting systems.

Video Transects

VIDEO CAN BE USED FOR MAPPING LARGE AREAS, and produces a permanent record of the community. A diver can swim along a transect holding the camera at a predetermined distance and angle to the bottom. The use of a transect tape allows a known area to be filmed and mapped.

For determination of benthic organism abundance, video photography can allow for the collection of a large amount of data in a short amount of time, with the analysis done subsequently on the surface. Quadrats can be placed by hand along a transect and then recorded on video. However, there can be some variability in focal distance from the camera to the substrate which can affect focus, resolution, and exposure. Alternatively, a video camera can be mounted to a "quadrapod," (see page 211) which essentially moves the quadrat with the camera, allowing for precise determination of focus and exposure (Coyer and Witman, 1990). Jaap et al. (1990) use a video camera mounted on a wire track, with the suspended camera pushed along the cable (transect) by a diver. In the laboratory, the video can be displayed as single frames, and scored by digitizing or using random dot patterns for percent cover and density of organisms.

Video transects can be used to determine fish density as well (see Chapter 7 and Training Exercise #12). Transects can be run on a predetermined swath size, or run as timed transects. There are quite a few studies that have evaluated some of the biases and problems with these type of fish surveys (Alevizon and Brooks,

FIGURE 8.10:
Video transect with camera and lights for benthic abundance estimation.
(photo courtesy of A. Gelber, PBS&J Co.)

1975; Sale and Douglas, 1981; Sale and Sharp, 1983; Bortone et al., 1986; Lincoln-Smith, 1988; Davis and Anderson, 1989). Some potential difficulties with this method include problems associated with water clarity, light levels, effects of observer swimming speed, species avoidance, and species identification (see Ebeling et al., 1980).

Larson and DeMartini (1984) used cinetransects to estimate the abundance of fishes on the bottom and in the water column. They used super-8 movie cameras, and did 3 minute transects where divers swam predetermined compass headings and photographed fish seen in a 120° horizontal about the transect axis and 1.5 m above and below the diver's depth. They calculated a transect volume, and expressed fish density per 1000 m³. They also evaluated the effects of underwater visibility on fish density estimations.

Scientific divers used quantitative video surveys in northern Canada to determine the spatial distribution and size of hypoxic "black pools of death" (to invertebrates and fishes; Kvitek et al., 1998). Video transects were run at a speed of 0.8 m/sec at 1 m above the seafloor, using a Sony Hi-8 camera mounted on a Dacor SeaSprint scooter. Percent cover of the black pools was estimated by dividing the length of time a point on the video frame crossed pools by the total run time. They also used digital frame grabs imported into an image analysis program (NIH Image v. 1.58) to calculate the area of each pool.

Video transects were used in the Aleutian Islands, Alaska, for determining benthic habitat descriptions and quantitative data on the distribution and abundance of major taxa (Jewett et al., 2008). A swath of 1 m by 25 m was videotaped for benthic species, and a 4 m by 25 m strip transect was used for a quantitative fish census.

An analysis of point quadrat and video transect techniques found that more taxa were found by the point quadrat method, and that the percent cover of rock was overestimated by video transects (Leonard and Clark, 1993). Video sampling required less time in the field, but more time in analysis in the laboratory.

Video has been used to study fish behavior such as social foraging. Barber and Auster (2009) used video to analyze mixed species foraging on the Great Barrier Reef. They used miniDV format videotape which was viewed on a VCR with shuttle search capabilities and a time code reader. Foraging events lasted an average of 1.35 min.

TRAINING EXERCISE #12
Quantitative Videography: Fish Transects

> **Equipment needed:** transect tape, underwater video camera
> **Objective:** To quantitatively evaluate the abundance of fishes

Ideally, there will be one video camera and transect tape per diving buddy group, but if not available, the group can share the equipment that is available. Prior to the dive, the staff should assign to each group random compass bearings for laying out the transect tape. Each group can use the video camera for a pre-determined timed transect, for example, five minutes, so that each group will have an opportunity to film. While one person is filming, the dive buddy can be taking visual census notes for comparison. The diver should record species and numbers of fishes seen in the same volume of water that will be analyzed on video.

Back in the laboratory, the video can be analyzed by each group for total number and species of fish seen on the transects. Comparisons should be made with visual estimates. Advantages, disadvantages, and limitations of each method can be discussed.

Time-lapse

VIDEO CAMERAS CAN ALSO BE PROGRAMMED for time-lapse photography and left in place underwater for an extended period of time. This can be useful for documenting the behavior of organisms, for looking at changes related to time of day or season, or for studying physical processes such as currents, surge, and scour.

Complex behavioral interactions, such as reproduction in fishes, can be analyzed by using video records played at slow, or single-frame speeds. Videographic techniques and direct diver observations of reproductive activity in pelagic spawning reef fishes have been shown to both be valid and comparable (Clavijo and Lindquist, 1992). Quantifiable spawning behaviors included spawning rushes, possible rushes, and false rushes.

Paquette et al. (2009) used remote underwater video to study species composition, behavior, and interactions of fish under close to natural conditions. They collected long time series imagery from single sites using six wide angle video cameras mounted at the center of a 15 m diameter circle. Time lapse was set up to take a 10 second clip once an hour. They were able to collect data on species richness, trophic guild composition, and species occurrence rates.

One system uses a movie camera and strobe unit housed in separate waterproof housings and mounted to a heavy pipe-frame tripod (Kendall and Bright, 1989). They set the camera to take one frame every 24 seconds, so that a roll of movie film would last for 24 hours. They assessed conspicuous diel cycles of reef fishes and invertebrates that may not otherwise have been possible by direct observation.

Time-lapse movie or video has been used to measure the frequency and duration of turtle feeding in grass beds (Ogden et al., 1980), kyphosid fish feeding (Littler et al., 1983), and limpet feeding (Kitting, 1979; Steneck, 1982). Other researchers have measured feeding events of fishes by counting the number of bites per unit time from video or direct observation (Kingett and Choat, 1981). Marrack (1999) used a Sony TR400 video camera in an underwater housing to record 5-minute intervals every hour for 24 hours in order to evaluate the movement of rhodoliths (non-geniculate corraline algae, see Chapter 2).

A variety of dyes can be released to track water flow directions and rates. See Maney et al. (1990) for further details on time-lapse videography and image analysis.

Stereo-video

Underwater stereo-video systems have been used for census assessments of reef fish abundance and length estimates of fishes (Harvey et al., 2004). These measurements have been found to be more accurate than diver visual estimates (Harvey et al., 2004). Remote systems can be used to avoid diver-disturbance, and be baited to attract fish. One diver-operated stereo-video system used two Sony TRV 900 video cameras in underwater housings mounted 0.7 m apart on a base bar (Watson et al., 2010). These two methods of assessing reef fish assemblage structure were shown to give very different results when used on the same transects and environment. A complete description of the system can be found in Harvey and Shortis (1996) and in Shortis et al., (2009).

FIGURE 8.11:
Diver swimming stereo-video system on fish transect in western Australia.
(photo courtesy of E. Harvey)

Websites

http://finepix.com/en/lagoon/ (Fuji underwater photography)

http://www.backscatter.com/ (underwater photography and videography equipment)

www.wetpixel.com

www.divephotoguide.com

www.uwphotographyguide.com

EQUIPMENT MANUFACTURERS AND SUPPLIERS

Video and photography suppliers

Amphibico (underwater video housings and lights)
459 Deslauriers
Montreal, QC, Canada
H4N 1W2
(514) 444-8666
www.amphibico.com

Backscatter (underwater photo and video)
225 Cannery Row
Monterey, CA 93940
(831) 645-1082
http://www.backscatter.com/

Canon USA Inc. (cameras, housings)
One Canon Plaza
Lake Success, NY 11042
(800) OKCANON
www.usa.canon.com

EWA Marine (underwater housings)
48 8171 4185-0
www.ewa-marine.com

Fantasea (cameras, housings, strobes)
2125 Collingwood St.
Vancouver, B.C.
V6R3K8 Canada
44 2921 250140
www.fantasea.com

Fuji (cameras)
www.fujifilm.com

Gates Underwater Products (cameras, videos, housings)
13685 Stowe Drive, suite A
Poway, CA 92064
(800) 875-1052
www.gateshousings.com

Helix (photography and video equipment, housings, strobes)
310 South Racine Ave.
Chicago, IL 60607
(312) 421-6000
www.helixcamera.com

Ikelite Underwater Systems (waterproof housings for cameras, video, strobes, lights)
50 West 33rd St.
Box 58100
Indianapolis, IN 46208
(317) 923-4523; (800) IKELITE
www.ikelite.com

Light and Motion (underwater video housings, lights, batteries)
300 Cannery Row
Monterey, CA 93940
(408) 645-1525
www.lightandmotion.com

Nauticam (underwater housings, port systems)
Rm 2203, CCT Telecom Bldg.
11 Wo Shing St.
Fotan, Hong Kong
www.nauticam.com

Nikon Inc. (underwater cameras, lens, strobes, accessories)
623 Stewart Ave.
Garden City, NY 11530
(516) 547-4200
www.nikon.com

NIMAR (underwater camera and video housings and lighting systems)
Via Pignedou, 8
42015 CORREGGIO (RE) Italy
Tel 39-0522-633026
www.nimar.it

Sea & Sea (underwater cameras, lenses, strobes, housings)
2105 Camino Vida Roble, Suite L
Carlsbad, CA 92009
(800) 732-7977
www.seaandsea.com

Seacam USA (housings, strobes)
www.seacamusa.com

SeaLife Cameras (underwater cameras and strobes)
Pioneer Research
97 Foster Road, suite 5
Moorestown, NJ 08057
(856) 866-9191
www.sealife-cameras.com

Sony Marine Pack (video cameras and housings)
www.sonystyle.com

SUBAL USA (underwater camera housings; port systems)
www.subalusa.com

References

Airoldi, L. and F. Cinelli. 1997. Effects of sedimentation on subtidal macroalgal assemblages: An experimental study from a Mediterranean rocky shore. J. Exp. Mar. Biol. Ecol. 215: 269-288.

Barans C.A., M.D. Arendt, T. Moore, and D. Schmidt. 2005. Remote video revisited: A visual technique for conducting long term monitoring of reef fishes on the continental shelf. Mar. Tech. Soc. J. 39(2): 110-8.

Barber, K. and P.J. Auster. 2005. Patterns of mixed-species foraging and the role of goatfish as a focal species. In: Godfrey, J.M. and S.E. Shumway (eds.), Diving for Science 2005, Proceedings of the Amer. Acad. of Underwater Sci., pp. 109-113.

Barsky, S.M., L. Milbrand, and M. Thurlow. Underwater digital video made easy. Hammerhead Press, Ventura, CA. 192 pp.

Blair, S.M. and B.S. Flynn. 1989. Biological monitoring of hard bottom reef communities off Dade County Florida: Community description. In: Lang, M.A. and W.C. Jaap, (eds.), Diving for Science...1989, Proceedings of Amer. Acad. Underwater Sci. pp. 9-24.

Cassidy, P.M. 1991. An inexpensive, rapid means of epibenthic community analysis. In: Krock, H.J. and D.E. Harper, Jr., (eds.), International Pacifica Scientific Diving...1991, Proceedings of Amer. Acad. Underwater Sci. pp. 1-5.

Christie, H. 1983. Use of video in remote studies of rocky subtidal interactions. Sarsia 68: 191-194.

Coyer, J., and J. Witman. 1990. The Underwater Catalog: A Guide to Methods in Underwater Research. Shoals Marine Laboratory, Cornell University, NY. 72 pp.

Coyer, J., D. Steller and J. Witman. 1999. The Underwater Catalog: a guide to methods in underwater research. 2nd Edition, Shoals Marine Laboratory, Cornell Univ., Ithaca, NY. 151 pp.

Curtis, A. Quantitative photography. In: Engle C, (ed.), Photography for the scientist. London, England: Academic Press, 1968: 438-512.

Davis, G.E. and T.W. Anderson. 1989. Population estimates of four kelp forest fishes and an evaluation of three *in situ* assessment techniques. Bull. Mar. Sci. 44:1138-1151.

Dayton, P.K., G.A. Robilliard, R.T. Paine, and L.B. Dayton. 1974. Biological accommodation in the benthic community at McMurdo Sound, Antarctica. Ecol. Monogr. 44: 105-128.

de Strobel, F. 1985. Underwater photography by a diving scientist: A useful combination of observations and documentation to design and apply oceanographic instruments. In: Mitchell, C.T. (ed.), Diving for Science...85, Proceedings of Joint International Scientific Diving Symposium, Amer. Acad. Underwater Sci. pp. 318-330.

Done, T. 1981. Photogrammetry in coral reef ecology: A technique for the study of change in coral communities. Proc. 4th Int. Coral Reef Symp., Manila. 2: 315-320.

Ebeling, A.W., R.J. Larson, W.S. Alevizon, and R.N. Bray. 1980. Annual variability of reef-fish assemblages in kelp forests off Santa Barbara, California. Fish. Bull., U.S. 78: 361-377.

Foster, M.S. 1975. Algal succession in a *Macrocystis pyrifera* forest. Mar. Biol. 32: 313-329.

Goldberg, N.A. and M.S. Foster. 2002. Settlement and post-settlement processes limit the abundance of the geniculate coralline alga *Calliarthron* on subtidal walls. J. Exp. Mar. Biol. Ecol. 278:31-45.

Hall, H. 1982. Guide to successful underwater photography: a new approach. Marcor Publishing, Port Hueneme, Calif. 189 pp.

Hanisak, M.D., S.M. Blair, and J.K. Reed. 1989. Use of photogrammetric techniques to monitor coral reef recovery following a major ship grounding. In: Lang, M.A. and W.C. Jaap, (eds.), Diving for Science...1989, Proceedings of Amer. Acad. Underwater Sci. pp. 119-135.

Harvey, E.S. and Shortis, M.R., 1996. A system for stereo-video measurement of sub-tidal organisms. Marine Technology Society Journal, 29(4): 10-22.

Harvey, E.S. and Shortis, M.R., 1998. Calibration stability of an underwater stereo-video system: Implications for measurement accuracy and precision. Marine Technology Society Journal, 32(2): 3-17.

Harvey, E., D. Fletcher, M.R. Shortis, and G.A. Kendrick. 2004. A comparison of underwater visual distance estimates made by scuba divers and a stereo-video system: Implications for underwater visual census of reef fish abundance. Mar. and Freshw. Res. 55: 573-580.

De Couet, H.G. and A. Green. The manual of underwater photography. Best Publishing, Flagstaff, AZ.

Jaap, W.C. 1986. A photogrametric apparatus for rapid areal benthic surveys. In: Mitchell, C.T. (ed.), Diving for Science...86, Proceedings of Amer. Acad. Underwater Sci. pp. 227-231.

Jaap, W.C., J.L. Wheaton, and K.B. Donnelly. 1990. Materials and methods to establish multipurpose, sustained, ecological research stations on coral reefs at Dry Tortugas. In: W.C. Jaap, (ed.), Diving for Science...1990, Proceedings of Amer. Acad. Underwater Sci. pp. 193-203.

Jaap, W.C., and M.D. McField. 2001. Video sampling for monitoring coral reef benthos. Bull. Biol. Soc. Wash. 10: 269-273.

Jackson, R.M. Essentials of Underwater Photography. Best Publishing Co., Flagstaff, AZ.

Jewett, S., R. Brewer, H. Chenelot, R. Clark, D. Dasher, S. Harper, and M. Hoberg. 2008. Scuba techniques for the Alaska monitoring and assessment program (AKMAP) of the Aleutian Islands, Alaska. Brueggeman, P. and N.W. Pollock, (eds.), Diving for Science 2008, Proceedings of Amer. Acad. Underwater Sci. pp. 71-89.

Johnston, C.A., I.S. Morrison, and K. MacLachlan. 1969. A photographic method for recording the underwater distribution of marine benthic organisms. J. Ecol. 57: 453-459.

Kendall, J.J. and T.J. Bright. 1989. An analysis of biotic interactions on the east Flower Garden Bank (Gulf of Mexico) using short-term time-lapse photography. In: Lang, M.A. and W.C. Jaap, (eds.), Diving for Science...1989, Proceedings of Amer. Acad. Underwater Sci. pp. 175-190.

Kingett, P.D. and J.H. Choat. 1981. Analysis of density and distribution patterns in *Chrysophrys auratus* (Pisces, Sparidae) within a reef environment: An experimental approach. Mar. Ecol. Progr. Ser. 5: 283-290.

Kitting, C.L. 1979. The use of feeding noises to determine the algal foods being consumed by intertidal molluscs. Oecologia 40: 1-17.

Kvitek, R.G., K.E. Conlan, and P.J. Iampietro. 1998. Black pools of death: Hypoxic, brine-filled ice gouge depressions become lethal traps for benthic organisms in a shallow Arctic embayment. Mar. Ecol. Progr. Ser. 162: 1-10.

Larson, R.J. and E.E. DeMartini. 1984. Abundance and vertical distribution of fishes in a cobble-bottom kelp forest off San Onofre, California. Fish. Bull 82: 37-53.

Leonard, G.H. and R.P. Clark. 1993. Point quadrat versus video transect estimates of the cover of benthic red algae. Mar. Ecol. Progr. Ser. 101: 203-208.

Lincoln-Smith, M.P. 1989. Improving multispecies rocky reef fish censuses by counting different groups of species using different procedures. Env. Biol. Fish. 26: 29-37.

Littler, M.M., P.R. Taylor, and D.S. Littler. 1983. Algal resistance to herbivory on a Caribbean barrier reef. Coral Reefs 2: 111-118.

Littler, M.M. and D.S. Littler. 1985. Nondestructive sampling. In: M.M. Littler and D.S. Littler (eds.), Handbook of phycological methods. Ecological Field Methods: Macroalgae. Cambridge University Press. New York. pp. 161-175.

Lundalv, T. 1971. Quantitative studies on rocky-bottom biocoenoses by underwater photogrammetry. A methodological study. Thalassia Jugoslavica 7: 205-213.

Lundalv, T. A stereophotographic method for quantitative studies on rocky-bottom Biocoenoses. In: Drew, E.A., J.N. Lythgoe, and J.D. Woods, (eds.), Underwater Research. 1976. Academic Press, New York. pp. 299-302.

Lundalv, T. 1986. Detection of long-term trends in rocky sublittoral communities: Representativeness of fixed sites. In: Moore, P.G. and R. Seed (eds.), The ecology of rocky coasts. Hodder and Stoghton, Kent, England. pp. 329-345.

Maney, E.J., J. Ayers, K.P. Sebens, and J.D. Witman. 1990. Quantitative techniques for underwater video photography. In: W.C. Jaap, (ed.), Diving for Science...1990, Proc. of Amer. Acad. Underwater Sci. 1990. pp. 255-265.

Marrack, E.C. 1999. The relationship between water motion and living rhodolith beds in the Southwestern Gulf of California, Mexico. Palaios 14: 159-171.

Ogden, J.C., S. Tighe, and S. Miller. 1980. Grazing of seagrasses by large herbivores in the Caribbean. Amer. Zool. 20: 949 (abstract).

Paquette, A.E., P.J. Auster, and M.D. Arendt. 2009. Approaches for analyzing behavioral interactions of fishes using time series video observations at an ocean observatory off the coast of Georgia. In: Pollock, N.W. (ed.), Diving for Science 2009, Proceedings of the Amer. Acad. Of Underwater Sci., Sea Grant Publ. No. CTSG-10-09, pp. 206-215.

Reed, S.A. 1980. Sampling and transecting techniques on tropical reef substrates. In: Environmental survey techniques for coastal water assessment conference proceedings. University of Hawaii Sea Grant College Program, Cooperative Report UNIHI-SEA-GRANT-CR-80-01. pp. 71-89

Sale, P.G. and W.A. Douglas. 1981. Precision and accuracy of visual census techniques for fish assemblages on coral reef patch reefs. Env. Biol. Fish. 6: 333-340.

Sale, P.G. and B.J. Sharp. 1983. Correction for bias in visual transect censuses of coral reef fishes. Coral Reefs 2: 37-42.

Sebens, K.P. 1986. Spatial relationships among encrusting marine organisms in the New England subtidal zone. Ecol. Monogr. 56: 73-96.

Sebens, K.P., E.J. Maney, Jr., and A. Gordon. 1997. Long term research in the rocky subtidal zone (Massachusetts 1977-1997). In: Maney, Jr. E.J. and C.H. Ellis, Jr. (eds.), Diving for Science...1997, Proceedings of Amer. Acad. Underwater Sci. pp. 141-159.

Shortis, M., E. Harvey and D. Abdo. 2009. A review of underwater stereo-image measurement for marine biology and ecology applications. Oceanogr. and Mar. Biol.: an Annual Review 47: 257-292.

Steneck, R.S. 1982. A limpet-coralline alga association: adaptations and defenses between a selective herbivore and its prey. Ecology 63: 507-522.

Taylor, H. 1977. Underwater with the Nikonos & Nikon systems. American Photographic Book Publishing Co., Inc. 160 pp.

Vroom, P.S., C.A. Musburger, S.W. Cooper, J.E. Maragos, K.N. Page-Albins, and M.A.V. Timmers. 2010. Marine biological community baselines in unimpacted tropical ecosystems: Spatial and temporal analysis of reefs at Howland and Baker Islands. Biodivers. Conserv. 19: 797-812.

Watson, D.L., E.S. Harvey, B.M. Fitzpatrick, T.J. Langlois and G. Shedrawi. 2010. Assessing reef fish assemblage structure: How do different stereo-video techniques compare? Mar. Biol. 157: 1237-1250.

Witman, J.D. 1985. Refuges, biological disturbance, and rocky subtidal community structure in New England. Ecol. Monogr. 55: 421-445.

Wu, N. 1994. How to photograph underwater. Stackpole Books, Mechanicsburg, PA.

INDEX

^{14}C method, 133

A
Aandera Instruments, 120
abalone, 139, 163
abundance, 139
Acoustic Doppler Current
    Profilers (ADCPs), 107
acoustic receiver, 178
acoustic tags, 167
acoustic transmitters, 178
ACSA Underwater GPS, 79
air-driven drill, 70, 71
airlift, 89, 159
algae, 153, 162
algae, marine, 134
algal frond growth, 136
American Academy of
    Underwater Sciences
    (AAUS), 4, 5, 7, 10, 12, 14,
    46, 54, 60, 183
American Nitrox Divers
    International (ANDI),
    60
Amphibico, 219
anchialine, 26
anchor ice, 24
anesthetize, 156
Antarctic, 23
AquaMap, 76
aquarium, 182
AQUARIUS Undersea
    Laboratory, 4, 50, 171,
    172

Aquatic Research
    Instruments, 184
aquifers, 27
archeology, 83
Arctic, 23
artifacts, 83, 90
Association of Diving
    Contractors (ADC), 7
Australian Scientific Divers
    Association (ASDA), 14

B
Backscatter, 219
bacteria, 150
band transect, 140, 146
baseline, 85
Baseline Survey Protocol
    (BSP), 143
beam transmissometer, 113
Bee Works, The, 184
behavior, 179
benthic, 143
benthic chambers, 116
benthic invertebrates, 160
benthic landers, 135
benthic organisms, 139
benthic respiration
    chambers, 135
benthic swath data sheet,
    144
benthos, 120
Benzocaine, 157
Betadine, 178

Biospherical Instruments,
    Inc., 120
bird, 178, 181
blue crab, 178
blue water, 24, 149
blue water diving, 24, 43
box core, box corer, 98
British Antarctic Survey
    (BAS), 23
Brownie's Third Lung, 58
Buddy Phone, 133
buoys, 69
BYNCKE net, 156

C
cage, 170
caging, 170
calcareous algae, 169
caliper, 167
camcorder, 214
Canadian Association for
    Underwater Science
    (CAUS), 5, 14
Canon U.S.A. Inc., 219
Carter Lift Bag, Inc., 58
caverns, 26, 54
caves, 26, 54
cement, 72
chemical measurements,
    116
chemical sterilization,
    174
Chicago Pneumatic, 79
climate change, 20

INDEX **227**

Coastal Leasing, Inc., 120
clod-cards, 105
closed circuit scuba, 47
clove oil, 157
Cochran Undersea
  Technology, 58
cold water diving, 44
collecting, 149
collection bags, 160
collection tubes, 152
color, 109
commercial diving, 50, 54
communications, 40
community stability, 23
compass, 74, 100
competition, 20
conductivity, 118
Confederation Mondiale
  des Activities
  Subaquatiques (CMAS),
  7, 14
CONSHELF, 3
contaminated water, 39, 40
coral, 162, 163, 172
coral bleaching, 20
coral reefs, 19, 20, 143
cores, 154
core sampler, 99
conductivity, 118
conductivity, temperature,
  and depth (CTD), 103,
  104
corer, 151
crab, 163
crayfish, 154
crepuscular behavior, 179
ctenophores, 180
current meters, 103, 105,
  106, 107

D

damselfish, 158, 165, 168
damselfish, dusky, 180
data, 132

data collection, 131
data loggers, 104, 137
data sheets, 74, 131
datum, 85
DCB, 12
Deep Ocean Engineering,
  79
demersal zooplankton, 152
Department of the Interior,
  5
depth certification, 12
depth gauge, 74
Desert Star Systems, 76, 79
destructive sampling, 143
diffusion samplers, 116
digital, 207
digital cameras, 208
digital SLR cameras, 209
digital flow meters, 106
digital video cameras, 214
digital videography, 214
dip measurements, 102
dip net, 155
disease, 138
dissolved oxygen, 117, 118
distribution, 139
dive log, 12
Dive Rite, 58
diver propulsion vehicle
  (DPV), 51
Divers Alert Network
  (DAN), 60
diversity, 20
DiveTracker Sport, 76
diving bell, 50
Diving Control Board
  (DCB), 6, 10, 12
Diving Equipment and
  Marketing Association
  (DEMA), 7, 58
diving helmets, 38
diving safety manual, 13
diving safety officer (DSO),
  10

diving sled, 28
Diving Unlimited
  International, 58
dolphin, 181
down line, 43
Draeger Safety, Inc., 58
dredge, 89
drilling, 70
dry suit, 39, 44

E

EAN 32, 45
EAN 36, 45
EarTag Company, The, 186
East Carolina University, 90
EdgeTech, 79
Efcom, Inc., 59
Ekman Grabs, 98, 152
Electrofishing, 157
El Niño, 116
EL Niño, effects of, 23
emergence traps, 153
Environmental Protection
  Agency (EPA), 7
epoxy, 72, 171
estuary, 19, 26
EWA Marine, 219
excavation, 89

F

fallout, 100
Falmouth Scientific, Inc.,
  120
Fantasea, 220
Farallon USA, 59
Federal Bureau of
  Investigation (FBI), 7
fence, 171
fin clipping, 167
fish, 145, 154, 164, 169, 177
fish aggregating devices
  (FADs), 177
fish behavior, 20
fish ecology, 20

fish populations, 23
fish sound, 114
fissures, 102
flow meters, 103
Floy Co., 156
Floy tag, 163, 165, 184
flume, 136
fluorescence, 111
fluorescent dye, 106
fluoricine dye, 175
fluorometer, 112
Forestry Suppliers, Inc., 79, 184
fossil, 99
freeze-branding, 167
freshwater, 26
Fuji, 220
full face masks, 38, 40, 133
fyke net, 156

G

gas-flow measurements, 119
gastropods, 163
Gates Underwater Products 220
gelatinous zooplankton, 149
General Oceanics, Inc., 120
Geographic Information System (GIS), 76, 132
geothermal, 29
geothermal spring, 103
Global Positioning System (GPS), 69, 76
goby, 166, 176
gravel suction pump, 100
grazer, 171
groundwater discharge, 102
guide number, 209

H

gand-fanning, 89
Hass, Hans, 47
Hawaiian sling, 158

HD video, 208
HD movie, 209
heat branding, 167
Heinz Walz, 184
Helix, 220
heliox, 45, 46, 49
Helle Engineering, 120
herbivory, 20
HID, 55
Historical Diving Society, The, 14
hookah, 40
hoop net, 155
horizontal offset, 85
hot springs, 29
HYDROLAB, 4
hydrophone, 113
hydrothermal vents, 29
hypercapnea, 48
hyperoxia, 48
hypersaline environments, 29

I

ice diving, 44
ice scour, 24
Ikelite Underwater Systems, 220
impact hammer, 70
inclinometer, 74, 75, 100, 101
infauna, 151
International Association of Nitrox and Technical Divers (IANTD), 60
InterOcean Systems, Inc., 121
Interspiro, 59
invertebrates, 162
ION, 79
irradiance, 109
irrigation pump, 160

J

JBL Enterprises, 184

J.W. Fishers Mfg., Inc., 80

K

kelp, 171
kelp forest, 19, 22
Kirby Morgan Dive Systems, 59
Klein, 77
Klein Associates, Inc., 80
Koppers, Inc., 80
krill, 180

L

lakes, 28, 29
Laminaria, 22
larvae, 152
laser, 169
LCD, 208
LED, 55
lenses, 208
LI-COR, Inc., 121
lift bags, 53, 54, 98
lift net, 155
light, 109
Light and Motion, 220
light meters, 103, 109
line-point intercept transect method, 142
lithics, 100
lobster, 163, 178
lock-out diving, 50

M

Macrocystis, 22
Macrocystis pyrifera, 135
macroinvertebrates, 154
magnetic stir bars, 134
magnetometer, 83, 84
mammals, 159, 181
mangrove forests, 20
map, 85
mapping, 74, 86
marine reserves, 23
marine snow, 150

marking, 69
measuring, 167
Memphis Net & Twine Co., 184
metal detectors, 78
Minerals Management Service (MMS), 6
mixed gas diving, 45
MOBY, 115
mollusks, 162
moorings, 115
Moss Landing Marine Laboratories, 115, 182
MS-222, 157
mussels, 172, 176

N

Nalge Co. Thermo Fisher Scientific, 185
Nalgene Polypaper, 131
NASA, 47, 115
National Association for Cave Diving, 60
National Band and Tag Co., 185
National Biological Service, 5
National Marine Sanctuary, 92
National Park Service, 5, 89
National Science Foundation (NSF), 5, 7, 23
National Speleological Society Cave Diving Section, 61
National Undersea Laboratory, 4
National Undersea Research Program (NURP), 4, 45, 50, 61
National Oceanic and Atmospheric Administration (NOAA), 4, 5, 7, 14, 45, 50, 61, 115
NAUI Worldwide, 61
Nauticam, 220
navigation, 69
navigate, 80
Nichols Net and Twine Co., 185
night diving, 55
Nikon, Inc., 221
Nikonos, 207
NIMAR, 221
NiMH, 55
nitrogen, 116
nitrogen narcosis, 48
nitrox, 45, 49
NOAA National Undersea Research Program (NURP), 14
nutrient, 116

O

Occupational Safety and Health Administration (OSHA), 5, 7, 12
Ocean Technology Systems, 59, 133
octopus, 163
offshore platforms, 25
Onset Computer Corporation, 121
open circuit scuba diving, 37
otoliths, marking of, 167
Outer Banks Outfitters, 185
outplantation, 171
Overton's, 185
oxygen, 45
oxygen microelectrodes, 117

P

PADI, 61
PAM, 112
PAM fluorometer, 138
parrotfish, 165
pelagic, 24, 145, 149
penetrometer, 98
percent cover, 141
percent cover, visual estimations of, 141
permanent transects, 140
piezometers, 102, 116, 117
photography, 86
photomosaic, 87, 88, 91
photosynthetically active radiation (PAR), 109
pH, 118
photodocumentation, 207
photogrammetric, 207, 211
photogrammetry, 213
photography, 207
photo quadrats, 143
photo-transects, 211
pH probes, 118
physiology, 137
Pingers, 77
PISCO, 140, 142, 147
piston cores, 98
plankton net, 155
point-intercept methods, 142
pole spear, 157, 164
Polychaetes, 164
Posidonia oceanica, 135
primary productivity, 133
Progressive Epoxy Polymers, 185
PVC, 141, 176, 177

Q

quadrapod, 211
quadrat, 139, 141, 142
quantitative photography, 211
quantum sensors, 109
quinaldine, 156

R

Ramset Fastening Systems, 185
random point contact (RPC), 142
rebreathers, 47, 48, 114
recordkeeping, 131
Recreation Supply Co., 80
recruitment, 173
red snapper, 166
Reef Check California, 147
reefs, artificial, 175, 176
reentry traps, 153
regulators, 38
Research Vessel Operator's Council (RVOC), 6
respiration chamber, 137
respiration measurements, 133
rhodolith, 21
Rite in the Rain, 131, 185
rivers, 28
RJE International, 80
rockfish, 164, 178, 179
rocky reefs, 22
rotenone, 157
ROV, 50, 53
Roving Diver Technique, 148
RPM Nautical Foundation, 91
rugosity, 100

S

safety diver, 43
salinity, 118
sand anchors, 103
sand ripples, 98, 100
sandy beach, 19
saturation diving, 2, 49
scallops, 163
scientific diver, 13, 37
scientific diving, 13
scientific diving certifications, 12
Scripps Institution of Oceanography, 2, 4
Sea & Sea, 221
sea anemones, 164
Sea-Bird Electronics, 121
Seacam U.S.A., 221
seal, 178
SeaLife Cameras, 221
sea lion, 181
sea otter/urchin/kelp interactions, 23
sea urchin, 164, 172
sea urchin grazing, 22-23,
seagrass, 21, 76, 169
SEALAB, 3
sea star, 164
secchi disk, 113
Secret Service, 7
sediment, 98
sediment collection, 98
sediment oxygen demand, 137
sediment traps, 100, 102
settling plates, 174
shark, 167, 183
shear strength, 102
side-scan sonar, 77, 83, 84
site survey, 84
size frequency, 167
Skidaway Institute of Oceanography, 116
slate, 74, 131
slave, 210
Smithsonian Institution, 2, 5
sonar, 69
SonTek/YSI, 121
Sony Marine Pack, 221
sound, 113
spectrometer, 111
sponges, 162

stationary visual survey, 146
Steam Machines, Inc., 59
stereographic photography, 212
stereo-video, 170, 218
strike measurements, 102
SUBAL U.S.A., 221
submarine canyon, 19, 25
Submerged Resources Center, 5, 89
Subsalve U.S.A., 122
substratum, 101
suction dredge, 159
suction gun, 154
suction sampler, 159, 161
surface-supplied diving, 39, 133
surge, 106
survey, 84
surveying, 86
syringe, 150

T

tag, 165
tagging, 90, 162
technical diving, 45, 46
Technical Diving International, 61
TEKTITE, 3
telemetry, 177
telephoto, 210
temperature, 104, 118
temperature logger, 117
tethered diving, 42
thermometers, 103, 104
through-the-lens (TTL), 209
tide gauge, 103
timed swimming transects, 145
timed transects, 146
time-lapse, 217
tools, 97, 131, 174

topographical survey, 84
Torpedo, Inc., 59
towboard, 52
towed-diver, 51
towed sled, 51
tow fish, 77
Training Exercises, 55
Training Exercise #1: Introduction to Specialized Equipment and Techniques, 57
Training Exercise #2: Locating and Marking a Site, 72
Training Exercise #3: Practice Using Tools Underwater (Multi-Station Exercise), 73
Training Exercise #4: General Underwater Mapping, 76
Training Exercise #5: Archaeology Underwater Mapping in Pool, 88
Training Exercise #6: Geological Mapping, 101
Training Exercise #7: Measurement of Physical Oceanography Parameters, 115
Training Exercise #8: Sample Survey of Common Benthic Biota-Familiarization with Transect Tapes, Quadrats, And Slates, 144
Training Exercise #9: Fish Transects, 149
Training Exercise #10: Collecting & Tagging Techniques, 167
Training Exercise #11: Photogrammetric Transect for Benthic Population Estimations, 212
Training Exercise #12: Quantitative Videography-Fish Transects, 217
transducers, 113
transect tape, 74, 139, 140
transplantation, 171
traps, 156
trawl diving, 180
Trelleborg Viking, 60
triangulation, 85
trimix, 45, 46
True North Technologies Corp., 80

U

underwater camera housings, 210
underwater habitats, 2, 49
Underwater Hyperbaric Medical Society (UHMS), 7
underwater magnifying device, 153
underwater mapping, 84
underwater marking, 70
underwater navigation, 74
underwater paper, 74, 131
underwater tape recorders, 133
uniform point contact (UPC), 141
United States Forestry Service, 7
United States Geological Survey (USGS), 7
University-National Oceanographic Laboratory System (UNOLS), 5, 7, 14
University of Michigan, 4
University of Washington, 4
U.S. Navy, 2, 7, 181
U.S. Navy Diving Manual, 7

V

van Dorn bottles, 116
VEMCO, 122, 178, 186
vertical offset, 85
vertical offset measurements, 84
Vexar, 170
video cameras, 217
videography, 86, 207
video lights, 215
video transects, 145, 215
vinyl tags, 162
visual census techniques, 145, 148

W

wall net, 155
Walz, 122
water induction dredge, 89
water motion, 105
waterproof housing, 209
wave buoy, 108
wave meters, 108
West Marine, 80, 186
Wet Labs, Inc., 122
whale shark, 182
wide angle, 210
Wisconsin Historical Society, 86
Woods Hole Oceanographic Institution, 86, 115
Wrasse, 179

Y

YSI, 122, 186

Z

zero visibility, 28
zero visibility conditions, 43
Z Spar marine epoxy, 72